EVERYTHING YOU NEVER WANTED TO KNOW ABOUT ERECTILE DYSFUNCTION AND PENILE IMPLANTS

End Your Silence, Sadness, Suffering, and Shame

RICK REDNER MSW
BRENDA REDNER RN

ISBN: 978-1-4834-5390-3 (sc)
ISBN: 978-1-4834-5389-7 (e)

Library of Congress Control Number: 2016910084

Lulu Publishing Services rev. date: 06/22/2016

CONTENTS

DEDICATION

I dedicate this book to my wife, Brenda, who remained by my side during the years I was impotent and felt undeserving of her love. I'm grateful our marriage vows included the promise to stay together "in sickness and in health." At the time we made that pledge to each other, we were clueless and naïve about the challenges ahead that would test our commitment to our wedding vows. Despite my best efforts to push her away, Brenda fought valiantly to preserve and protect our marriage. I'm blessed to have her devotion, dedication, loyalty, and love. Brenda, I love you with all my heart.

I also want to express my gratitude toward the doctors, nurses, and all of the healthcare professionals who dedicate years of their lives to learning how to treat and heal all varieties of illness, injury, and disease.

Additionally, I'd like to thank Dr. John Payne for his willingness to spend many hours reading the rough drafts and commenting and encouraging me along the path to making this book a reality.

Last but not least, I'd like to thank the soverign God of the universe who gave me the opportunity to share this journey with you.

INTRODUCTION

My wife, Brenda, and I are reluctant authors. We never imagined we'd be writing a book about our experiences with erectile dysfunction (ED) and penile implant surgery. Erectile dysfunction and impotence are not issues we've felt comfortable discussing with our family or our closest friends, but in these pages, we will address these issues candidly with you. We do it in the hope that it will help you to discuss these issues with your doctor, your partner, and, most importantly, with yourself. How you talk to yourself about ED affects your emotional health, your relationship with your partner, and your sexuality. Therefore, it's important to learn healthy ways to speak to yourself about this disease. As Brenda and I share our lives and our struggles with you, our goal is to provide you with the information, tools, motivation, and courage to deal with erectile dysfunction in a healthy and constructive way.

In my lifetime, I've experienced two traumatic and life-changing conversations with my doctors. To this day, I can't decide which of those conversations was worse. In the first conversation, I was told I had prostate cancer. The second conversation occurred approximately four years after my bilateral nerve-sparing prostate surgery. In that conversation, I was told the healing process for my nerve bundles was complete. Since I was unresponsive to penile injections and ED medications, I would be impotent for the rest of my life. Both of those conversations ushered in a new era in my life that began with me feeling emotionally devastated.

Ironically, the urologist who gave me the dreadful news was the same person who performed my penile implant surgery. It remains a mystery to me why he did not mention the option of an implant during that first visit. I suffered needlessly for months before I came up with the idea of a penile implant on my own. Far too many men like me are left feeling miserable, hopeless, depressed, and unaware of the amazing alternatives to restore erectile function.[1] Whatever is causing your erectile dysfunction, hope exists, but determining the best way to treat your erectile dysfunction involves an act of courage.

The journey toward healing begins with the potentially embarrassing task of making an appointment to see a doctor to evaluate the cause of your erectile dysfunction. The Mayo Clinic lists the following potential medical and psychological causes of ED:[2]

- Heart disease
- Clogged blood vessels (atherosclerosis)
- High cholesterol
- High blood pressure
- Diabetes
- Obesity
- Metabolic syndrome—a condition involving increased blood pressure, high insulin levels, body fat around the waist, and high cholesterol
- Parkinson's disease
- Multiple sclerosis
- Peyronie's disease—development of scar tissue inside the penis
- Certain prescription medications
- Tobacco use
- Alcoholism and other forms of substance abuse
- Sleep disorders
- Treatments for prostate cancer or enlarged prostate
- Depression, anxiety, or other mental health conditions
- Stress
- Surgeries or injuries that affect the pelvic area or spinal cord
- Relationship problems due to stress, poor communication, or other concerns

Take a careful look at the list above. If you've been unable or unwilling to contact a physician, it's time to examine what you've been telling yourself about your erectile dysfunction. It's time to challenge your excuses, embarrassment, or fear and take the important step of making that appointment. It's the first step on the journey toward healing.

CHAPTER 1

Losing Your Erectile Abilities: A Traumatic Life-changing Event

If you are coping with erectile dysfunction you are not alone. According to the National Institutes of Health (NIH), erectile dysfunction affects as many as 30 million men in the United States.[1] While ED occurs more frequently as men age, it affects young men as well.

My journey with erectile dysfunction began after I was diagnosed with prostate cancer at age fifty-seven. I chose robotic surgery to treat the cancer for two reasons. First, I wanted every cancer cell in my body removed surgically. Second, I was told that with this type of surgery, if both my nerves were spared, there was no doubt I would experience a return of my erectile function.

Later in my journey I discovered the information I had received was inaccurate. According to the National Cancer Institute, of the 219,000 men in the United States who learn they have prostate cancer each year, nearly half undergo surgical removal of the gland. If I had read the article "Regrets After Prostate Surgery," I would have known that: "Overall, the vast majority of men were satisfied. However, nineteen percent regretted their treatment choice. Notably, men who had undergone robotic surgery were four times more likely to regret their choice than men who had undergone the open procedure. Part of the problem may be that doctors who perform robotic prostatectomies commonly cite potency rates as high as ninety-five percent and above among their patients, giving patients an unrealistic view of life after surgery."[2]

When I was told my cancer was confined to my prostate and that my surgeon considered me cured, I was elated. In my wildest imagination I never thought that within thirty days of receiving the news I was cancer free that I would be so depressed I would have fleeting thoughts of killing myself. Losing both urinary control and erectile function was too much for me to cope with. I hated the quality of my life without a prostate. I was sorry I had survived the surgery.

Most men and couples facing life without a prostate do not receive the help or preparation they need for the changes that prostate surgery brings about. My wife and I wrote our first book, *I Left My Prostate In San Francisco—Where's Yours?*[3] to help men and couples cope with the situation in which we found ourselves following my surgery.

One issue that leads frequently to depression is the loss of erectile function. ED has a direct impact on a man's life and marriage. "It is suggested that ED is involved in one in five failed marriages."[4]

The majority of men spend some time struggling with depression as a result of erectile dysfunction. When men are depressed, they tend to mask the symptoms. Men who are depressed rarely appear sad. Instead, they become agitated and easily angered. They might experience a loss in energy, lose interest in activities they enjoyed previously, and withdraw from people. Depression can also lead to a noticeable reduction in desire and/or interest in sex. To medicate themselves, many men turn to alcohol or substance abuse.

Unfortunately, men are far less adept at recognizing their symptoms than women. A man is more likely to deny his feelings, hide them from himself and others, or try to mask them with other behaviors. The three most common signs of depression in men are:

1. Physical pain. Sometimes depression in men shows up as physical symptoms—such as backache, frequent headaches, sleep problems, sexual dysfunction, or digestive disorders—that don't respond to normal treatment.
2. Anger. This could range from irritability, sensitivity to criticism, or a loss of your sense of humor to road rage, a short temper, or even violence. Some men become abusive, controlling, verbally or physically abusive to wives, children, or other loved ones.
3. Reckless behavior. A man suffering from depression may start exhibiting escapist or risky behavior. This could mean pursuing dangerous sports, driving recklessly, or engaging in unsafe sex. You might drink too much, abuse drugs, or gamble compulsively.[5]

As a result of his depression, a man's relationship with his partner might be overwhelmed with fighting and negativity. Men coping with erectile dysfunction are easily trapped in negative thinking about themselves, their sexuality, and their primary love relationship. The cycle of negative thinking, combined with irritability, fighting, and emotional and sexual withdrawal, can stress any relationship to the breaking point.

In addition, far too many men avoid seeking medical or psychological help as they cope with erectile dysfunction. Powerful feelings of shame and embarrassment fuel their resistance to seek professional help. It can be a stressful and difficult time. If you are in a relationship, the odds are both of you will be unhappy. If you are single, ED might inflict a fatal blow to your self-esteem. Your confidence inside and outside of the bedroom can plummet to the extent you withdraw from dating and forego all thoughts of marriage.

My ongoing experience with erectile dysfunction led me to feel angry. I hated living without my prostate. I wanted to go back in time to change my treatment decision. I wished I had kept my cancerous prostate. My reaction was not uncommon. Almost all men who have had their prostate removed experience some form of remorse. For some men, the feeling is temporary, for other's it's a lifelong feeling of regret. An article in *Men's Health* entitled "I Want My Prostate Back" does an excellent job of expressing this sense of remorse.[6]

Against my will, I joined the ranks of men who suffered from post-surgery remorse. No one prepared me to face life, love, and sexuality without my prostate. I hated living without the ability to achieve an erection. Like most men coping with ED, I began my journey in total ignorance without help, information, or support. I felt unworthy as a man. That led me to believe Brenda would be better off without me. As a result of being depressed and feeling unworthy as a man and a husband, I pushed Brenda away by ignoring her emotional and physical needs. I hoped my pushing her away would make her decide to leave me. I believed my wife deserved to be with a man whose sexuality wasn't damaged by cancer and surgery. I thought no woman in her right mind would want a relationship with man whose manhood had been taken away.

Thankfully, Brenda took her wedding vows seriously. When she pledged to stay by me in "sickness and in health" she meant it. For her, "sickness" included losing my erectile abilities. As a result of her steadfast love, I discovered that embracing a new sexuality while living with erectile dysfunction takes time, intentionality, flexibility, creativity, a willingness to try new things, and lots of discussions with your partner.

ED is more than a man's disease; it affects the man's partner as well. According to an article in *Sexual Health Australia*, the partners of men with erectile dysfunction feel:[7]

- Guilty about being unable to make their partner's penis erect
- Helpless and unsure where to go for support
- Rejected/unloved, not sure if their partner is still into them
- Unattractive
- Confused/frustrated
- Scared/worried that he is having an affair
- Worried about the health and wellbeing of their loved ones
- Insecure about his love

- Sad for him, or even for herself, due to not having an erect penis in the bedroom anymore
- Relieved their man's penis is not all they bring into the bedroom anymore

In addition to the stress ED creates in relationships, the effects of ED leak into other parts of a man's world. "ED affects not only the relationship with his partner but may also affect how the man interacts with friends and co-workers. A man may lose his confidence, his enjoyment in life, and morale. Outwardly, a man might project a macho image but inside may not feel like he measures up."[8]

Whether you are single or married, it's important for you to know and believe that if you have ED, you are not doomed to a life without enjoyable sex. You aren't disqualified from finding a lifelong partner. I learned that it's possible to experience a mutually exciting and satisfying sex life without erections, and so can you. What I'm about to share in this book might be difficult to accept, but I hope you'll believe it's possible.

The first step to achieving a mutually exciting and satisfying sex life living with ED involves confronting the belief that you are no longer a man because you are incapable of achieving an erection. If you believe ED robbed you of your manhood, that belief will rob you of the motivation necessary to discover the ways you and your partner can nurture and maintain a healthy, exciting, and mutually satisfying sex life.

We need to deal with one more important issue before we move on. Men lose more than their erectile function while coping with ED. They also lose their sexual triggers, which we'll address in the next chapter.

Questions to consider:

1. Do you know the cause of your erectile dysfunction?
2. Have you consulted with a physician?
3. How has erectile dysfunction affected your sense of manhood?
4. How has it affected your relationship with your partner?

CHAPTER 2

ED Is a Thief

Before a man develops erectile dysfunction, he has a number of exciting sexual triggers that result in an erection. Some of the pre-ED triggers for an erection include:

- **Sight**: Men often get aroused as a result of seeing something that turns them on.
- **Sounds**: Certain sounds or sex talk can arouse men. An entire industry is based on the power of sex talk to arouse men.
- **Smells**: Particular smells can cause a man to experience an erection.
- **Fantasy**: Men have the capacity to use their imagination to think about sexual experiences that arouse them, resulting in erection.
- **Words**: Reading about sexual encounters can be highly arousing.
- **Touch**: Certain forms of touching can create a state of arousal.

Erectile dysfunction robs you of your capacity to respond with an erection to each and every one of your previously exciting sexual triggers. The situation gets worse with the passage of time. After living with ED for a relatively short period of time, all of your previous erection-producing triggers elicit painful emotions, such as disappointment, frustration, anger, and shame. To protect themselves from experiencing these powerfully unpleasant feelings, men will avoid any actions or behaviors that are associated with their previous sexual triggers. That's one reason why most men struggling with ED withdraw from all forms of physical affection. Holding hands, backrubs, kissing, and all other forms of affection become bitter reminders of what has been lost.

The majority of couples are bewildered and confused when they discover their entire sexual history is lost to both of them. Few couples seek help in grieving these losses and establishing a new sexual relationship that does not depend upon their history with previous sexual triggers.

Reclaiming your sexuality post-ED takes time, effort, experimentation, and new experiences to re-program your mind and body to experience arousal in completely different ways than you did when your erectile function was intact. I've heard from far too many men who, in the face of losing their erectile function and sexual triggers, give up on their sexuality for the remainder of their life.

It's important for every man living with ED to understand that it's possible to find pleasure, enjoyment, and orgasms with a flaccid penis. However, it isn't a simple, easy, or quick lesson for a man to learn. For decades, a flaccid penis was associated with the absence of desire. Now a man must learn how to feel excitement while remaining flaccid. It is an essential lesson to learn in order to reclaim and renew your sexuality without an erection. I suspect it takes a complete rewiring of the brain for a man to learn that he can be aroused sexually with a flaccid penis.

To maintain your sexual triggers while coping with ED, I suggest you pay close attention to the ways in which you speak to yourself now that your favorite triggers no longer produce erections. The odds are you'll hear yourself speaking in harsh, judgmental tones. You become your harshest critic, thinking things like you are no longer worthy of being loved by anyone. All of us maintain a committee inside our head that evaluates our performance and provides us with continuous feedback and commentary about our behavior. When it comes to coping with ED, you need to identify and fire those committee members whose voices harshly condemn or mercilessly criticize your sexual prowess. Replace those committee members with new ones who possess the ability to speak kindly and lovingly yet have the ability to speak both grace and truth in a balanced way. This enables you to speak to yourself in comforting ways as you acknowledge and then grieve the losses you've experienced since the onset of erectile dysfunction.

It's also helpful if your partner has the ability to support you by speaking loving and encouraging words to you. Remember, you are not the only one facing unwanted loss and change. Your partner might be grieving as well. It will be difficult, challenging, and emotionally painful, but it's also part of the healing process if you are able to support your partner through her grieving process. Far too many couples are unable or unwilling to grieve their losses together, which leaves both of you to grieve your losses alone.

Based on my discussions with a lot of men, I discovered I wasn't the only man who didn't know that it is possible to experience an enjoyable orgasm with a flaccid penis. Even though you might be unable to maintain an erection, it is possible for you to re-claim your pre-erectile dysfunction triggers by creating new and pleasurable associations with those triggers that end with you experiencing an orgasm.

For example, if a certain smell was an erection-producing trigger, make sure you have orgasms in the presence of that smell many times until that trigger gives you new and pleasurable associations. Repeat the process with all of your sexual triggers. By doing so, you will build new associations of sexual excitement and orgasms in the presence of your favorite sexual triggers. Over time, with or without an implant, reclaiming your favorite sexual triggers is an important part of the healing process. During the process, it's highly likely you'll discover new triggers that are both pleasurable and exciting for you and your partner. Add those newfound triggers to your sexual life together.

There are three highly traumatic and vulnerable times in life of an impotent man and couple living with ED. The first occurs at the moment of impact, when erectile dysfunction takes away a man's ability to have an erection. The second occurs in response to grieving the loss of sexual triggers. Many men experience high levels of depression during this time.

They withdraw from their partner both physically and emotionally. They also experience an increase in relational tension and fighting. The third point of vulnerability comes from how men respond to coping with their loss and depression.

Some men learn from adversity. They discover ways to enjoy their sexuality despite ED. The men who give up on their sexuality face the most psychological and relational challenges. Most men become so preoccupied and devastated with their loss that it does not occur to them that their partner is hurting as well. To make matters worse, they feel so bad about themselves they don't want to hear how ED is affecting their partner. Most men avoid discussing their ED at all costs.

Four years into my journey with ED, I was stuck. There was nothing left to offer that would give me the ability to achieve an erection. I wanted to read a book written about penile implants. As I searched for books, I couldn't find a single one that contained the information I needed to help decide whether or not to go through with implant surgery. For me, that was a defining moment. I knew it was important for my wife and I to share our experiences so other couples could learn to cope successfully with ED or get the information they needed about penile implants so they could decide whether or not to move forward with implant surgery.

If you've decided to do what's necessary to restore your erectile function, the first step is to determine the cause of your erectile dysfunction, because it has a direct bearing on your treatment options. For example, for men whose erectile dysfunction is a result of psychological issues or prescription medication, implant surgery is the wrong treatment option.

Therefore, before you do anything radical, consult with your doctor and/ or a specialist to find the underlying cause of your ED. For some men, counseling or a change in prescription medication is all that's needed to cure their ED. In addition, most insurance companies will not pay for a penile implant unless there is a documented medical cause for erectile dysfunction. Unless you plan to pay for your surgery privately, there is no way to skip this step in the journey toward implant surgery.

By the time you are finished reading about our journey and the information contained in this book, you'll be better prepared to make a decision about whether a penile implant is the right decision for you. If you decide to pursue this option, you'll have our tips and suggestions to spare you the mistakes and unnecessary suffering that my wife and I endured. Welcome aboard. Sit back, and relax into your own personal circumstances as we take a journey together, which might end with you regaining your erectile abilities.

Questions to consider:

1. What were your preferred/favorite sexual triggers?
2. What are your partner's preferred sexual triggers?
3. How has losing your sexual triggers affected your desire for sex?
4. Are you still engaged in pleasing your partner or have you given up on your and your partner's sexuality?
5. Ask your partner to share her thoughts about discovering the ways to enjoy both her and your sexuality.

CHAPTER 3

How You Think About ED Changes Everything

Discovering I had prostate cancer was devastating, but when I received the news that my ED was permanent, I took a second emotional nosedive. That's when I discovered such nosedives or low points are also relevant to men coping with erectile dysfunction.

For the last four years, I've been reaching out via the internet to thousands of men with prostate cancer. During that time, I've observed three distinct orientations toward cancer and erectile dysfunction. 1) the Diver, 2) the Survivor, 3) and the Thriver. Earl Lynch was the first to use these classifications in his article about transforming a customer's experience.[1] I use the classifications to identify the different phases of coping with ED.

The Diver Phase

The majority of men diagnosed with prostate cancer and erectile dysfunction will spend some time in the Diver phase. How do you know you are in the process of taking a nosedive? You will experience a few of the following symptoms:

- Trouble sleeping or concentrating
- Persistent and intrusive thoughts about cancer or erectile dysfunction

- Ever-present feelings of anxiety, fear, and irritability
- Feelings of helplessness or loss control
- A sense of pessimism or fear about your future
- Excessive worry, where you think and believe you will suffer all kinds of worst-case scenarios
- Depression, isolation, and withdrawal from previously satisfying relationships
- Drug or alcohol abuse to alter your mood or help you to cope
- Suicidal thoughts

If you're stuck the Diver phase and have had fleeting thoughts of suicide or you've actually come up with a plan, please read the rest of this chapter carefully as if your life depended upon it, because it might.

After my wife and I wrote *I Left My Prostate in San Francisco—Where's Yours?* I developed a number of online and social media sites to reach out to men and couples coping with prostate cancer. Through one of those sites, I received a letter that replays in my mind frequently. It convinced me of the importance of writing this book for men and couples struggling with ED. Some of the facts have been changed to protect the identity of this couple, but here's the letter:

> *"My husband had a robotic prostatectomy. It cured his cancer, but left him impotent. This killed his soul. We loved each other deeply and each of us knew that, but there was nothing I could do to heal his pain. If I tried to initiate intimacy, he would become anxious and push me away. Alternately, he would initiate intimacy when he had self-medicated with alcohol. This was difficult for me and never had a good outcome. Humor didn't go far, either. He felt damaged. He wasn't a group kind of guy, so he never received professional help for his emotional pain. He went deeper into depression. On September 10th he committed suicide. While his impotence wasn't the only issue that drove him to his decision to end his life, it was a major factor in his feeling life wasn't "worth living." It's heartbreaking for our family."*

Here's a man who was diagnosed and cured of prostate cancer. The price he paid for his cure was the loss of his erectile function. His refusal to seek help left him to cope with his emotional and relational pain alone. He isolated himself from his wife and used alcohol to self-medicate. When his pain reached overwhelming levels, he decided to take his life. He left behind a loving and heartbroken wife.

As a loner who did not seek outside information or support, I suspect he died as a result of ignorance. It's highly doubtful he knew that implant surgery was available as an option to restore his erectile function. Brenda and I wrote this book to provide men and couples an alternative to the hopelessness and despair brought about by erectile dysfunction.

To every man who is stuck in the Diver orientation with thoughts of suicide, I understand your desire to make your suffering go away. I understand the pain of losing your sense of manhood, because I've been there. I want you to know that you have more value than an erect penis. I urge to discuss your thoughts of suicide with more than one person. Here are some people who can support you during this difficult phase in your life:

- Your partner
- Your family physician
- A friend
- A professional counselor
- Your pastor, priest, rabbi, or imam
- An online or face-to-face support group

I don't believe manhood demands that we remain the strong, silent type who deals with emotionally devastating issues alone. I believe it takes a man to admit when he needs help and to ask for that help. If you are one of those men who refuses to ask for driving directions when lost, you will probably find it even more difficult to ask for help now that you're experiencing ED. Being a man means being willing to do what's difficult.

Thoughts of suicide are the equivalent of the "check engine" light in your car. When that red warning light goes on, you do something constructive. If you can fix the problem by yourself, you will. If you can't fix the problem yourself, you don't pull out a hammer and destroy your car. Since you value your car, you take your car to a mechanic who has the skills to diagnose and then fix the problem. If you are depressed or having suicidal thoughts, something is seriously wrong. It's up to you to choose how you'll respond to the red warning light flashing in your life. Speaking as one man to another, I urge you to do the right thing. In the same way you bring your car to a mechanic, professionals are available to help you cope with or cure your erectile dysfunction.

The National Institute of Mental Health[2] identifies the following risk factors for suicide:

- Depression and other mental disorders
- Alcohol or substance abuse
- A prior suicide attempt
- Family history of a mental disorder or substance abuse
- Family history of suicide
- Family violence, including physical or sexual abuse
- Having guns or other firearms in the home
- Incarceration, being in prison or jail
- Being exposed to others' suicidal behavior, such as that of family members, peers, or media figures

If you or your partner have fleeting thoughts of suicide, the "check engine" light in your life is flashing. If you've developed a specific lethal plan, and/or there is a gun in the home, you've moved way past the red warning light. Immediate help is required. Call 911. If the threat is not immediate, two additional resources are available seven days a week, twenty-four hours a day.

- The National Suicide Prevention Lifeline: 1-800-273-8255
- The Veterans Crisis Line: 1-800-273-8255 (press 1)

The first obstacle that many men need to overcome, whether it involves coping with depression, suicidal thoughts, or erectile dysfunction, is their resistance to seek help. By making a commitment to finish reading this book, you've added Brenda and me to your team. We appreciate that! By the time you finish this book, we believe you'll have gained something valuable that can make a huge difference in your life and in your attitude, beliefs, and emotional reactions to your erectile dysfunction. What is that thing? Hope. With the instillation of hope comes the belief that things will get better. If the restoration of your erectile function is vital to you, you'll know the next steps you need to take to restore your erectile function and put an end to your impotence, which, by the way, is an infinitely wiser decision than putting an end to your life. This is so important that I'll say it again: If you are determined to put an end to something, put an end to impotence rather than your life. Suicide is *not* the cure for erectile dysfunction.

If you are the partner of a man who refuses to seek help, you've probably asked yourself the following question dozens of times: "How can I help someone who doesn't want to be helped?" I think the best way to reach someone who is resistant to getting help is to start off with an option that takes minimal effort and is least threatening. Asking your partner to read this book or join an online community using an anonymous screen name are two options that fit that description.

Some men use anger and defensiveness to shut down discussion about their erectile dysfunction. If you've attempted to plead or argue with a man who does this, you know from experience that arguing is a waste of time and effort. For men who use anger and defensiveness, a different strategy is required. You need to create a situation where not dealing with their ED becomes more painful than dealing with their ED. For example, you might say something like, "I regret to inform you that as long as you continue to refuse to see your doctor about your ED, I'm sleeping in a separate bedroom." You don't need to state the consequence in a threatening or angry way. You need to say it in a calm, loving, and caring way. Outside of the bedroom, you can remain friendly, but keep the consequence in effect for as long as it takes to produce a change in attitude or behavior.

For your man, it could take more than one painful consequence to motivate him to overcome his embarrassment and shame and get the help he needs. Some men are more resistant to seeking outside help than others.

Other ideas to increase the pain of refusing to deal with ED are to stop cooking, shopping, or cleaning. Do whatever it takes to cause a significant disruption in your relationship until he takes action. Once again, it's important to remain friendly and kind as you apply this pressure. I hope the men who agree to read this book will not require this type of pressure to motivate them to deal with their ED in healthy ways and to seek the appropriate help.

If you are living with ED and reading this book, I want to congratulate you on your willingness to take in new information. A teachable spirit is necessary and required to cope successfully with erectile dysfunction. Brenda and I have experienced the emotional and relational damage that occurs when ED visits a couple. Our goal is to offer you an opportunity to heal your relationship emotionally, relationally, physically, and spiritually whether or not you decide to go with implant surgery.

The Diver orientation is the most emotionally, relationally, and sexually painful orientation with which to cope. It is difficult to engage in constructive problem solving when you are stuck in the Diver orientation. Remaining stuck in the Diver orientation is similar to discovering you have just traveled hundreds of miles in the wrong direction on a journey. It will take time and effort to get back on the road to healthy coping. It also takes time to form a team for direction and help. The Diver orientation is the most difficult orientation to endure. The two other orientations are not nearly as disruptive to your self-esteem, sexuality, emotional health, and your relationships.

I spent far too long in the Diver phase, because I was too ashamed to seek help. I suffered, and so did my relationship with Brenda. If you or your partner believe you are depressed or withdrawn, it's time to seek help. If you refuse, it's possible to get stuck in the Diver phase for years or decades. Take it from someone who spent far too long there: You don't want to live in that phase a day longer than necessary.

Survivors

Survivors place a high value on overcoming problems. They view ED as a problem to solve. They want to get over things without too much introspection. They have a high need for control over the events and challenges in their lives. They are assertive. They tend to have large networks of support, but they are not necessarily introspective. Survivors are conquerors. Many people in this group are socially and politically active. They are leaders. They start support groups and look for ways to help others. The best way to help an ED survivor is to offer jo, options for each problem they face. If the issue is coping with ED that is not responding to ED medication, offer jo, a vacuum pump or penile injections or the possibility of a penile implant. He'll make a decision. If he finds an option that works, it's on to the next problem that needs to be solved. Men in this group recognize that erectile dysfunction has put a lot of stress on their relationship with their sexual partner. Some will admit their relationship is not as good as it was prior to their diagnosis of ED. Yet they get up every day and do what needs to be done. Survivors take pride in overcoming obstacles, and their ED is one of those obstacles. If an ED survivor could push a button and go back to their pre-erectile dysfunction life, the overwhelming majority would do it without any hesitation. Men with a survivor orientation want solutions to the problem they face without much discussion. A woman married to someone with an ED survivor orientation might be frustrated that her partner doesn't place a high value on listening to or understanding the emotional or relational components of the affliction. He simply wants to solve the problem without much discussion.

Thrivers

Thrivers will tell you how their experience with ED has transformed their life as an individual and as a couple. They will tell you how much closer they feel toward their partner. They'll tell you they've become more loving, open, honest, and caring. They'll explain how their sex life isn't the same as it was prior to ED but how they have discovered new ways to satisfy each other.

Thrivers enjoy their physical and sexual relationship with their partner. Many will say their sex life is different and even better than their pre-ED days. They are a strong and closely-knit couple. They go places together and enjoy their lives, because they've learned how to how fun together. Laughter and humor plays an important part of their lives. They cultivate an attitude of gratitude. They are grateful for the smallest things, and they'll thank you profusely for any act of kindness you show toward them. If you ask them if they could push a button and go back to their pre-ED days of functioning and thinking, most will say "no."

Since personalities and reactions to impotence are so varied, the three orientations are not meant to describe every man's reaction to ED. Additionally, they are not meant to be sequential. The goal isn't for everyone to become a Thriver. Of the three orientations, the one you want to avoid is the Diver.

Questions to consider:

1. Which (if any) of the three orientations describe where are you in your journey?
2. If the Diver phase describes your current orientation, what steps will you take to move out of this orientation?
3. What is your long-term goal as you live with erectile dysfunction?
4. Discuss the ways in which you've resisted seeking help.
5. Discuss the ways in which shame, embarrassment, or fear keep you away from the resources that can help you.
6. Do you have support coping with ED? If not. are you willing to seek support?
7. How will you overcome your shame, embarrassment, or fear to enable you to seek help?

CHAPTER 4

What You Don't Know Can Hurt You

From a medical standpoint, a sudden onset of ED might be an early warning signal that could save your life. *"Eighty percent of men who land in the ER with a first heart attack say they developed ED at some point in the three years before,"*[1] says Daniel Shoskes, MD. He is a professor of urology at Cleveland Clinic. To protect your most precious asset, your health, you need to see a physician for a complete exam to determine the underlying cause of your erectile dysfunction. Don't allow depression, embarrassment, or shame to keep you from seeking a medical consult for ED (or any other disease or symptoms you experience). Men are usually resistant to utilizing our healthcare system. Sometimes I wonder if stubbornness is the leading cause of death for men in America.

What I'm about to share is an unfortunate reality. When you reach out to the medical community, you should expect little, if any, support or acknowledgement regarding the emotional, relational, psychological, or sexual difficulties you'll experience while coping with ED. Frankly, I find this reality maddening. It is well known in the medical field that men who receive a diagnosis of erectile dysfunction experience a catastrophic and devastating loss. They face a significantly higher risk for depression. They also suffer from dramatic loss of self-esteem. Most couples experience a significant increase in stress, conflict, and negativity. In every role a man performs, be it husband, father, or co-worker, he might experience a noticeable decline in his functioning. If all of this is known to the medical field, why are men who face a diagnosis of ED are not given any warnings or told to seek help if they experience relational problems or symptoms of depression? Why do men receive more information and warnings when they buy a pack of cigarettes than they do when they are diagnosed with erectile dysfunction?

Recall the letter I received from the widow whose husband killed himself. I doubt he was told there are a variety of ways to treat ED depending upon the cause of the disease. Here are a few of the options available:

- Counseling
- Muse (a penile suppository)
- A vacuum pump
- A variety of ED medications
- Penile injections
- A penile implant
- A change in medications

Here are some lifestyle changes that can improve erectile function:

- Losing Weight
- Quitting smoking
- Abstaining from alcohol
- Increasing physical activity
- Abstaining from illegal drugs
- Abstaining from pornography

A stubborn refusal to deal with erectile dysfunction is dangerous to your emotional, psychological, physical, sexual, and relational health. When it comes to ED, we have a chicken and egg problem. Which came first, your ED or depression? One leads to the other. Either way, depression should not be ignored.

When a woman is depressed, there is usually evidence of sadness. When a man is depressed, sadness might be absent. *Health* magazine[2] lists twelve symptoms of depression in men:

- Fatigue or loss of energy
- Sleeping too much or too little
- Physical complaints, such as a backache or stomach ache
- Irritability
- Difficulty concentrating
- A noticeable increase in anger and/or hostility
- Stress
- Anxiety
- Substance abuse
- Sexual dysfunction
- Indecision
- Suicidal thoughts
- Impotence [I added this one to the list]

If you have a history of depression or a family history of depression and/or suicide, you are at a much higher risk to remain stuck in a Diver orientation until you seek professional help. Depression is highly treatable, and yet, according to Health Line, it is estimated that two thirds of people living with depression never seek help.[3]

If you would like to gain a better understanding as to whether or not you are experiencing some of the symptoms of depression, visit this link HYPERLINK: http://www.webmd.com/depression/depression-assessment. If you decide to take this survey, keep in mind it is *not* designed to make a definitive diagnosis of depression or take the place of a professional diagnosis. Do not use the results of this test to avoid seeking a professional assessment. Depression saps your motivation and energy. It's difficult, perhaps impossible, to cope with erectile dysfunction in a constructive or healthy way when you are depressed.

Take the narrow path and seek help in the battle to defeat depression. Here are a few resources to help you win that battle:

1. Help Guide.Org:
   HYPERLINK "http://www.helpguide.org/articles/depression/depression-in-men.htm" http://www.helpguide.org/articles/depression/depression-in-men.htm

2. Psych Central:
   HYPERLINK "http://psychcentral.com/lib/where-to-get-help-for-depression/" http://psychcentral.com/lib/where-to-get-help-for-depression/

3. The Anxiety and Depression Association of America:
   HYPERLINK "http://treatment.adaa.org/" http://treatment.adaa.org/

Questions to Consider:

1. Are you or your partner coping with depression?
2. If you're aren't sure, will you take the depression survey?
3. If the survey indicates you are coping with depression, will you seek help?
4. If changes in your life style could improve your erectile functioning, are you willing to make those changes? If not, why not?

CHAPTER 5

"Danger, Will Robinson!"

As a kid, one of my favorite TV shows was *Lost in Space.* Anytime Will, the youngest son in the Robinson clan, was in danger, the Robinson's trusty robot would flail his arms up and down while he shouted, "Danger, Will Robinson!"

A few weeks before I was going to turn this manuscript in for publication, I heard from a man who had had his prostate removed recently. While he was coping with erectile dysfunction, he had made many sexual advances toward his wife. After multiple rejections, his wife finally told him the bitter truth. She wasn't interested in any type of sex until he was able to "get it up." I was grieved to hear that. I know that any time one partner rejects the other's sexuality, the relationship suffers unavoidable damage. If, by a miracle, her husband regained his erectile function the next day, the relational damage caused by that rejection could still have a negative impact on them for weeks or months to come. If he remains impotent for an extended period of time without a change in their style of coping, the couple is headed for a diminished level of cohesiveness and intimacy, with the possibility of divorce down the road.

When a couple marries, they make an emotional, relational, and sexual commitment to each other. It is a reasonable expectation to have your sexual needs satisfied in the context of your marriage. When one partner decides the sexual relationship is over, he or she is rejecting his or her partner's sexuality in such a way that he or she is placing the relationship in grave danger. This is true for unmarried couples as well.

If either one of you have rejected or withdrawn from the other emotionally, physically, or sexually as the result of coping with ED, I wish I could send the *Lost In Space* robot to your door to give you that familiar warning, "Danger, Will Robinson!" It would be nice if a simple admonition of, "Don't reject each another" set couples on a healthy path to discovering new ways to express their sexuality together, but it doesn't. It takes work to get there. Experiencing grief, loss, or depression can make it beyond your current ability to change your behavior on your own. As a member of your team, I feel obligated to warn you that making the wrong choices while coping with ED can destroy your relationship. I know it's difficult and/or embarrassing, but don't let your relationship become a casualty of ED. If either of you are experiencing difficulty coping with ED, both of you need to seek help together.

I could not find a centralized place for couples to call to get professional help for coping with ED, so I suggest you start with your physician. Many online groups are also available. I've included a few of them in the notes for this chapter on page XX.

Three important character traits will help you meet the challenge of coping with ED. The effectiveness of these character traits are greatly diminished by untreated depression. The first is *a teachable spirit.* Men with a teachable spirit learn from their mistakes, hardship, suffering, and life experiences. They suffer setbacks, but they don't let their setbacks define who they are. More importantly, they don't make the same destructive mistakes repeatedly. Do you possess a teachable spirit? If so, you will make good use of the material contained in this book to help you cope with ED. If not, you probably stopped reading this book long ago.

The second characteristic is *persistence.* Persistence is the willingness to persevere in the face of obstacles and hardship. Folks who are persistent will not be defeated or paralyzed by shame. They will take all the steps necessary to find out whether or not ED can be treated. They will not stop until they've exhausted every option. If it comes down to living with ED, persistent men and couples do not give up. They are determined to become the best lovers they can be.

The third important character trait is *flexibility*. Flexibility allows you to think outside the box. Men and couples who are flexible don't get stuck grieving the past. They are adaptable. They have a willingness to experiment with new ideas and options. They find creative, playful, and imaginative ways to pleasure each other and establish a satisfying sex life. Men and couples who have a teachable spirit and are persistent and flexible will find their way to a new and satisfying sex life with or without ED.

All couples who are coping with ED struggle for a period of time. It's more of a challenge if one or both of you lack a teachable spirit, persistence, or flexibility. It's an even greater challenge if your relationship was in serious trouble before you were confronted with the challenges of coping with ED. If that describes your situation, I believe professional help is necessary to save your relationship.

Another indicator of the need for outside help is if you or your partner lack in one of the three important traits. Your relational and sexual life will benefit from outside help if you learn to develop these character traits in your current circumstances. That help can come from a trusted friend, an online support group, or from a professional. There is no reason to spend the rest of your life in misery or give up on the pleasures that come from sex. With or without a cure for ED, your sex life isn't over unless you allow the hopelessness of depression to make that decision for you. I can summarize the message of this entire book in three words: *Don't give up!*

Questions to consider:

1. How can you and your partner develop a teachable spirit, persistence, and flexibility to help you cope with ED?
2. What other character traits do you need to help you cope with ED?

3. What is the current state of your relationship with your partner? (Ask him or her to discuss this with you.)
4. How can you make use of a teachable spirit, persistence, and/or flexibility to take the necessary steps to improve your situation?
5. What new and creative ways can you find to pleasure each other sexually?
6. Do you or your partner believe you need outside help? (Hint: If one of you says "yes," then both of you need outside help.)
7. Do you and your partner want to live with ED, or do you want to develop a plan of action to treat your ED?
8. Are both of you in agreement with that plan? (Hint: If one of you says "no," make another plan.)

CHAPTER 6

What's Faith Got to Do with It?

As a Christian author, I share my story in my books, blogs, and posts in social media. It's unavoidable for me to share how my faith impacts what I say, think, and do, because my faith is part of who I am. It's by and through our faith that Brenda and I decided to write this book. That said, we are aware that sharing our faith has the potential to offend readers who have a different faith or no faith. I'm giving you a heads-up that you might disagree or take offense at some of the material in this chapter.

If you are turned off by the topic of faith, you might be tempted to skip this chapter. Please don't give in to that temptation. If you do, you'll miss out on the important and useful information contained herein. I am confident that individuals and couples of any faith or of no faith will find useful and practical ideas in this chapter that can transform their experience with erectile dysfunction.

For example, this chapter provides you with the opportunity to learn your primary love language and your partner's love primary love language. Learning to speak each other's love language is an amazingly effective way to change your relationship for the better. As you begin reading this chapter, expect the opportunity for great things to happen in your relationship.

As you live out the day-to-day realities of coping with ED, the strength of your faith will be tested as it is whenever you are confronted with any life-changing illness, injury, or disease. For a large number of people, faith plays an important and meaningful role in how they cope with health-related struggles. In the furnace of adversity, some folks lose their faith, others find it, and some are like Brenda and me—their faith grows in direct proportion to the difficulty they are facing.

At the beginning of my journey, I was disappointed that my faith did not provide me with immediate relief from the emotional and relational pain as I coped with both my diagnoses of prostate cancer and erectile dysfunction. What I didn't understand was that my faith wasn't going to take my pain away or provide me with a miracle cure. Since I wasn't open to the possibility that ED was a pathway to discover many blessings, I missed out on more than a few of those blessings. I began my journey with ED thinking of it as a curse. I didn't believe anything positive could come from it. I believed all it offered my wife and me were misery, despair, and a defective sexuality, which I found impossible to enjoy. My attitude became a self-fulfilling prophecy for the first eighteen months of my journey with ED.

Looking back, I wish I had prayed for the wisdom that God wanted to provide to me about erectile dysfunction much earlier in my journey. I believed I prolonged my suffering. I made no effort to challenge my belief that losing my erectile abilities meant I'd lost my manhood. I was too angry, depressed, and miserable to accept any other reality.

It was through the lens of my faith that I challenged the belief that erections=manhood. Through my faith, I discovered this equation for manhood was a lie. Whether or not you believe that Jesus is Lord and Savior, you can study His teachings to gain an accurate perspective on what it means to be a man.

After I turned to Jesus and allowed His life and teaching to revolutionize the way I viewed manhood, I was freed from the culturally powerful but misguided belief that manhood depends upon the ability to achieve an erection. One of the many gifts of ED is that it offers you the opportunity to redefine what it means to be a man.

Jesus became my role model for manhood. He chose to live a life that did not include sexual pleasures, yet He remains the greatest example of what manhood is all about. His teaching and his life provided me with the greatest example of the embodiment of masculinity and love.

If you are truly interested in becoming an expert on love, the following passage in scripture will help you achieve your goal of knowing how to love your partner. I believe this is the greatest exposition about the nature of love. It's found in 1 Corinthians 13:

> Love suffers long and is kind; love does not envy; love does not parade itself, is not puffed up; does not behave rudely, does not seek its own, is not provoked, thinks no evil; does not rejoice in iniquity, but rejoices in the truth; bears all things, believes all things, hopes all things, endures all things. Love never fails.[1]

It's neither natural nor easy to love anyone in the ways described in these verses. Depending upon your experiences, your family of origin, and your current level of functioning, this might be impossible for you without the help of a mentor or counselor. Some folks have gaps or damage in their personal development or are in the midst of a personal crisis, which make it impossible for you to give or receive love in this way without receiving outside help.

Those who call Jesus their Lord and Savior have a special helper to make these traits bear fruit in your life. That helper is the Holy Spirit. The Holy Spirit offers you additional assistance with very special fruit. Galatians 5:22-23 says, "But the fruit of the Spirit is love, joy, peace, longsuffering, kindness, goodness, faithfulness, gentleness, self-control."[2]

Whether or not you believe in the biblical definition of manhood, you need to challenge the formula of erections=manhood and come to a different conclusion regarding what it means for you to be a man. Women also need to discover they can be attractive and desirable without the ability to give their man an erection.

When you learn to love in a sacrificial and serving way, you are on the path to true manhood. Your faith, or your lack thereof, will impact how you and your partner cope with ED.

It might seem out of place to include the importance of maintaining your sex life in a chapter about faith, but it's my belief that a mutually satisfying sex life is built into God's design of marriage. Therefore, giving up on your sex life as a result of ED results in losing all the benefits and blessings a mutually satisfying sex life brings to your relationship. Sadly, as a result of the powerful combination of ignorance and mistaken beliefs, Brenda and I experienced a season of losing those benefits and blessings. This occurred when I became convinced it was impossible for me to become a masterful student of both relational and sexual love without the ability to have an erection.

I've learned it takes time, effort, and an attitude adjustment on the part of both parties affected by ED for a couple to reestablish a new sex life that doesn't depend upon erections. It also takes teamwork, flexibility, a willingness to try new things, and a sense playfulness. If either of you clings to the belief that the only way to have a satisfying sexual experience is for the man to have an erection, then both of you will be miserable for as long as the man remains impotent. Both of you have the opportunity to take a huge step toward psychological, emotional, relational, and sexual healing when you look at that formula, challenge it, and conclude it is misguided, inaccurate, and an outright lie.

Romance and a mutually exciting sex life does not need to stop when a man loses his erectile function. Once you grieve the loss of your pre-ED sexuality, you can find enjoyment in your post-ED sexuality. As I mentioned, I had no idea could experience an orgasm with a flaccid penis. You can! Will it be the same? No. Will it be enjoyable? It can be if you let yourself enjoy your experience without comparing it to what you've lost.

Going through surgery for a penile implant is not the answer for everyone diagnosed with erectile dysfunction. If you've made the decision to live with your condition, I hope you'll do so with a willingness to embrace and enjoy your sexuality. Many books and articles discuss the various ways to enjoy sex without erections. Authors Ralph Alterowitz and his wife Barbara wrote *The Lovin' Ain't Over: The Couple's Guide to Better Sex After Prostate Disease*[3] to help couples reestablish a satisfying sex life while coping with impotence. Many other resources are available to couples who don't want to give up on the blessings of a satisfying sex life.

Re-establishing a sex life while living with ED requires a degree of flexibility and expanding what you are and are not willing to do. For example, some couples find oral sex very exciting. Others find the concept of oral sex disgusting. I suggest you consider overcoming your aversion to oral sex. It's a wonderful way to enjoy and satisfy each another once intercourse is no longer an option.

The issue of climacturia (leaking urine) before or after an orgasm can be a turn off for one or both of you. I was more repulsed than my wife as we dealt with the issue. For my peace of mind, we purchased a mattress protector. We also kept a towel on the bed. When it was my time to have an orgasm, I went to the bathroom first. It was neither romantic nor spontaneous, but for my peace of mind, it was necessary. I was delighted to discover I would leak significantly less urine if I lay on my back during my orgasm. Those were embarrassing issues we needed to resolve in order to reestablish our sex life together.

Once again, our faith and our belief that God designed marriage to include the blessing of a mutually satisfying sex life helped us to overcome the many obstacles we faced, including the challenge of discovering new ways to enjoy our sexuality while coping with erectile dysfunction following prostate surgery.

A key ingredient in finding a new, mutually satisfying sex life is to become comfortable talking about sex. Each of you needs the freedom to discuss and experiment with new activities to determine what you like and don't like. It might feel awkward and embarrassing to talk about and then experiment with novel activities, but the use of vibrators or other sex toys can enhance your enjoyment of sex.

Unless you are willing to read, learn, and discuss this together, you won't find mutually agreeable and satisfying ways to enhance your pleasure and enjoy your sexuality. As you discuss these issues, expect conflict and disagreement. That's not a bad thing as long as you respect each other's boundaries, thoughts, and opinions. Playfulness, flexibility, and a willingness to experiment with new and unfamiliar behaviors are necessary to establish a new sexuality together.

Choosing an implant means your time living with erectile dysfunction is temporary, but take it from me, you'll want to keep the pleasurable things you learn on this journey in your relationship once you regain your erectile function. Choose the path of learning to enjoy sex without an erection regardless of whether or not you decide to have implant surgery.

Here's a life-changing suggestion with the potential to increase your level of intimacy in your relationship. Gary Chapman[4] wrote a book in which he identified five primary love languages. They are:

1. Gift giving
2. Quality time
3. Words of affirmation
4. Acts of service
5. Physical touch

In his research, he found that many people gave their partner gifts of love based on their own preferred love language rather than their partner's favorite love language. This mistake causes couples to miss the expressions of love they give to each other. By taking this survey together and sharing the results with each other, you'll have the opportunity to learn how to give acts of love to your partner in his or her primary love languages.

Here's a link to take this survey:[5]

HYPERLINK "http://www.5lovelanguages.com/" http://www.5lovelanguages.com

After the survey, don't limit your expressions of love to one of the five languages: shower each other with all five, but pay particular attention to your partner's primary love language. If you do this, you'll experience a positive change in your relationship. If you are single and dating, at some point, you can suggest taking this survey together as a means of increasing your emotional connectedness with each other. If you are coping with ED, this survey is great news, because it shows you don't need an erection to speak any of the five love languages.

If you are stuck in negative thinking about your ED and/or your relationship is suffering, one non-threatening way for you to receive immediate support is for you to join an online support group. By choosing a screen name, you can keep your identity private. If you don't want to join, you can learn a lot from reading the posts from other men coping with ED. Here's a list of a few online erectile dysfunction forums:

Fanktalk ED Forum:[6]

HYPERLINK "http://www.franktalk.org/phpBB3/"

Daily Strength ED Forum:[7]

HYPERLINK "http://www.dailystrength.org/c/Impotence-Erectile-Dysfunction/forum" http://www.dailystrength.org/c/Impotence-Erectile-Dysfunction/forum

MedHelp ED Community:[8]

HYPERLINK "http://www.medhelp.org/forums/Erectile-Dysfunction/show/124"

Over time, I discovered that my faith had something far more valuable to offer to our journey than the immediate relief of pain. Our faith sustained us, comforted us, and enabled us to learn valuable life lessons. It served as the bond that kept us committed to our relationship. I suspect, though I can't be certain, that without our faith, our marriage would not have survived the last four years.

As I type this sentence, in a few hours, my wife and I will be heading off to the coast to celebrate our thirty-fifth wedding anniversary. These last four years of coping with cancer and ED have been filled with ups and downs, victories and defeats. We've witnessed God work in our lives and in our relationship. I can't imagine taking this journey without a belief in a God who loves us more than we can possible know or understand. If you began your journey with ED without any faith in God, perhaps your journey with ED will lead you to discover God.

During my years as a medical social worker, I met a number of people who were convinced that God gave them their cancer as a punishment for something awful they did in the past. If that were true, everyone on the planet would have cancer. It's not within the scope of this book to discuss why God allows suffering, disease, and evil. Yet, as you deal with erectile dysfunction or any other form of illness, injury, or disease, questioning God's goodness can result in a spiritual crisis.

If there is a spiritual issue that's important for you to resolve, taking action is an important part of your healing process. Consider speaking with someone from your faith tradition, such as a friend or a religious leader, who can help you resolve the spiritual aspects of coping with ED. Most crises will not leave your faith as they found it. You will grow closer to or further from God. It was my desire to grow closer. It's up to you to decide how coping with ED will affect your faith and how your faith will impact your ability to cope with ED.

Questions to consider:

1. How has your faith affected your attitude toward yourself and your partner as you've coped with ED?
2. How does your faith define manhood?
3. In what ways have you allowed ED to take away your manhood?
4. What blessings, gifts, or wisdom has coping with ED offered to you?
5. Have you prayed about your current circumstances, asking for wisdom?
6. Were your prayers answered?

CHAPTER 7

The Long and Winding Road Toward Surgery

This wasn't supposed to happen. Prior to my prostate surgery, I completed a pre-surgical sexual history. Based on the survey, my surgeon was optimistic that bilateral nerve-sparing surgery would preserve my erectile function. After surgery, when we were told both nerves were spared, we had every reason to believe I would recover my erectile function.

We were totally unprepared for the many ways that living without a prostate would affect my sexuality. I found the changes devastating, both emotionally and physically. Three months after surgery, I was still impotent. I found a video online titled "Penile Rehabilitation after Cancer Treatments."[1] I was shocked to learn that men who do not regain their functioning within the first three months can expect to wait somewhere between eighteen to twenty-four months for their nerve bundles to heal. As weeks turned into months, and the months into years, my wife and I experienced psychological, emotional, sexual, and relational trauma.

No help or information was provided to me as a cancer survivor facing impotence or to my wife. We were about to go through a major earthquake in our marriage blind, unaware, ignorant, and without any preparation or support. In my discussions with thousands of men following prostate surgery, few of them received any information before or after surgery as they struggled with the quality of life issues that prostate surgery brought on.

Like most men who chose surgery, moments after my catheter was pulled, I experienced severe loss of urinary control. I knew that would happen, but I had no idea how living in diapers 24/7 would impact me emotionally or sexually. I didn't know how often to change my diapers. After leaking urine through the diaper and onto my clothing a few times in public, I decided to avoid leaving the house. That's not all I avoided. I had no interest in sex. I did what most men do when avoiding sex: I stopped all expressions of affection with my wife. I avoided hugging, holding hands, and kissing. My penis became my enemy. I was so disgusted with the constant leaking of urine that I lost interest in maintaining a sexual relationship.

I began the journey of reclaiming my sexuality without a prostate in total and complete ignorance. No one from the medical field told me about a post-surgery issue that would totally disgust me and send me back into a deep, dark depression. I was furious when I read Dr. John P. Mulhall's book, *Saving Your Sex Life,*[2] where he explained that sexual stimulation results in the relaxation of the pelvic floor musculature. For men with ED, this means you'll leak urine during foreplay, and you might ejaculate several ounces of urine during climax. The issue, called climacturia, plagued me for years.

According to Dr. Mulhall, up to ninety percent of men who have their prostate removed experience this at least once, and twenty percent experience climacturia on a consistent basis for one to two years. I felt betrayed that no one had warned me about this issue prior to my surgery. After surgery, no one in the medical field offered any assistance with this common post-surgical issue, and I was too ashamed to ask.

The issue of climacturia was the last straw for me. At that point in my journey with cancer and impotence, I was ready to put an end to my sex life. I was prepared to live out the rest of my life as a eunuch. Brenda did not agree. She became a woman on a mission. She wanted me to do everything possible to regain my erectile abilities. I was too depressed and disgusted to care.

As a result of Brenda's daily prompting (which I experienced at the time as nagging), I participated reluctantly in a penile rehab program. The goal of penile rehab is to take steps to prevent venous leak. A venous leak prevents the penis from holding enough blood to maintain an erection. For men who've had their prostate removed, the goal of penile rehab[3] is to get oxygenated blood into the penis to prevent the onset of a venous leak, which results in erectile dysfunction. I began penile rehab by taking a small daily dose of a ED medication two weeks prior to surgery and continued to take the medication every day after surgery.

Following surgery, men are often encouraged to use a vacuum pump to obtain erections. Since the blood a vacuum pump brings to the penis is not highly oxygenated, there is some controversy as to whether the vacuum pump does anything to preserve erectile function. I found the vacuum pump to be very uncomfortable. Additionally, I could not get an erection hard enough for penetration. I decided the vacuum pump was useless. In less than a month, I tossed the very expensive pump into the garbage.

The next step in my penile rehabilitation was learning how to perform penile injections. They worked well for three months. After that, the injections mysteriously stopped working, even with the highest dose. I went in for a consult to try a different medication. The burning sensation I felt after the first injection was so severe, my days of penile injections came to an abrupt end. Both Brenda and I were joyful to say goodbye to that phase of my penile rehab.

After penile injections, I was placed on a variety of ED medications. I discovered that many ED medications have awful side effects, such as painful backaches or severe migraine headaches. It took a long time of experimenting with a variety of medications and dosages before we found a medication and a dosage that I could tolerate. I experienced intermittent success. Sometimes I'd achieve a useable erection, but more frequently, I wouldn't. It was the perfect formula to undermine my confidence in the bedroom.

On occasion, I experiences a nighttime erection, which led me to believe my nerve bundles were healing. Each time that happened, we developed a fresh hope that my erectile abilities would return. Then months went by before it occurred again. It was the most unpleasant roller coaster ride of our lives. We experienced a mixture of successes and failures with a variety of ED medications at different doses for three years. Brenda has vivid memories of our successful experiences. All I remember are the disappointing failures.

I wanted to try everything possible, so my next step was testosterone. I used testosterone in combination with ED medication. Despite increasing my dosage multiple times, I had no success. When the news broke about the high incidence of stroke associated with using testosterone, I stopping using it immediately. My family has a long history with strokes. I did not want to do anything that would increase that likelihood. Years later, other studies have contradicted and/or limited this finding regarding injectable testosterone. Therefore, it's important you discuss the risks and benefits of testosterone with your physician before you begin testosterone treatment.

Following the failure of multiple forms of penile rehabilitation, I made another appointment with my rehab urologist, expecting him to offer me yet another option to restore my erectile function. I didn't anticipate that my journey with penile rehabilitation was about to end.

During that fateful appointment, we were told the healing process for my nerve bundle was complete. There was nothing else he could offer to treat my erectile dysfunction. I was one of the unfortunate men who would not regain his erectile function following bilateral nerve-sparing prostate surgery. That meant I'd spend the rest of my life impotent. To this day, I can't understand why he didn't suggest the possibility of an implant during that office visit. Since he didn't offer that option, and I didn't know to ask, I left my appointment convinced I had just received a life sentence of impotence. I felt helpless and hopeless. I was deeply depressed before we reached the parking lot. We drove the ninety-mile journey home in complete silence.

I couldn't make sense of what had happened to me. I had been told I could expect a full return of erectile function. To ensure that outcome, I participated in a penile rehabilitation program and did everything I was told to do, but none of it mattered. The loss of my erectile function was the causality of my battle with prostate cancer. I concluded that the price I had paid to win that battle was too high. I developed a severe case of buyer's remorse, and then I got stuck in an "if only" loop.

- If only I had eaten less meat and dairy products, I wouldn't have had prostate cancer.
- If only I had refused to go for a biopsy, I wouldn't have discovered I had cancer.
- If only I had done more research on treating prostate cancer, I might have picked a different treatment option.
- If only I hadn't chosen surgery.
- If only I'd chosen active surveillance.
- If only I'd chosen the cyber knife.

Nothing productive came from my ruminations and self-recrimination. For months, I remained in a deep, dark depression. I wished for a way to go back in time, to reverse my decision to have prostate surgery, even if it meant dying of prostate cancer. It's easy to think that way if you've never seen anyone die from prostate cancer.

I'm sorry to say it took me far too long to realize it was time to take my life sentence of impotence to God in prayer. I didn't pray for a miracle healing. I prayed for peace and wisdom to make the best of my current circumstances. During one of my talks with God about my circumstances (which I refer to as prayer), I was reminded of an event that occurred three decades earlier.

As a medical social worker, I had met with a wheelchair-bound patient in the hospital who told me that he had a penile implant. I had no idea whether a penile implant would work for me, but I was determined to find out. I believe that recalling that patient and his experience with an implant was an answer to my prayers.

I went to Amazon to buy a few books about penile implants. I was shocked to discover I couldn't find a single book with the information I was seeking. Next, I searched for a book on the web. I couldn't find one there either. Suddenly, I had a déjà vu experience. I was disappointed with the lack of information available to help me make such an important, life-changing decision. I was certain that tens of thousands of men and couples were facing similar circumstances. Like me, they were living unhappily with erectile dysfunction and totally unaware that a penile implant was an option. That information gap was our motivation to write our first book about our experiences living without a prostate. Looking back, I believe it was also the defining moment that resulted in us writing this book.

For the moment, I put those thoughts aside. I needed to learn a lot more about penile implants. After giving up on the idea of finding a book, I decided to search for articles about penile implants. I had two goals. First, to learn the different between the types of implants that were available. My second goal was to join a forum for men who had undergone the surgery and were willing to share their experiences.

I learned that most men who choose implant surgery are like me in that they've tried and failed with a variety of treatment options before considering an implant. Experience with multiple failures brings about anger, frustration, hopelessness, and despair, both emotionally and relationally. Multiple failures can produce unhealthy self-esteem as well as an unhealthy sense of skepticism, which leads men to believe there isn't an effective treatment available that will cure their erectile dysfunction.

As I researched penile implants online, almost every article and study said the same thing: that penile implants are the most effective way to treat ED. Additionally, implants received the highest levels of patient and partner satisfaction. I was still skeptical though. Over the course of the last four years, all of my previous experiences treating ED had ended in the same way—failure. Was it possible I had stumbled onto the one treatment that could resolve my erectile dysfunction?

Before finalizing my decision about implant surgery, I joined an online penile implant forum called FrankTalk:

HYPERLINK "http://www.franktalk.org/phpBB3/viewforum.php?f=6" http://www.franktalk.org/phpBB3/viewforum.php?f=6. The forum is designated for discussions about penile implants. I began by asking men about their experiences with implants. Not one man who wrote back regretted his decision to have implant surgery. Those who had any regrets were the men who had waited and suffered with ED for years or decades. Those men expressed the regret that they had lived with ED far too long before having surgery. I didn't want to experience that regret.

From my perspective, I had already suffered long enough. I had waited four years hoping and expecting that my erectile abilities would return. That was long enough for me. The treatment of my prostate cancer had taken away my erectile function, and I wanted that back. There was no way I wanted to be impotent for the rest of my life. Between the studies and the firsthand testimonials about the success with implants, I had all the information I needed. I was certain a penile implant was the right way for me to end my experience with ED.

My wife was in Vermont visiting our oldest son when I made my decision. On a long-distance call, I announced my decision to have the surgery for a penile implant. I expected Brenda to share my excitement and enthusiasm. It was not the first or the last time in our marriage that I assumed wrongly. She wasn't pleased at all. I became defensive and angry. It seemed to me that she that she preferred me to live with impotence rather than go through a surgery to restore my erectile function.

Once Brenda returned home, I asked her to read the studies and the comments I had received from men who had the surgery. I was disappointed when Brenda finally agreed to support my desire to have penile implant surgery with a sense of dread rather than excitement. She had vivid memories of my history with surgeries. Almost every one of my previous surgeries was fraught with unexpected and serious problems. Brenda was certain not one but many things would go wrong. She was convinced I was heading down a road that would cause us more unexpected and unanticipated hardship, suffering, and grief.

I didn't want to listen to any of Brenda's concerns. I attributed her doubts to her tendency to worry unnecessarily. When she finally gave her blessing, she tried her best to warn me of the rough road ahead. I paid no attention. I was convinced it would be one of the easiest surgeries of my life. I was certain that, with the right surgeon, nothing of any consequence would go wrong.

My excitement about finding the solution to end my four-year journey with impotence blinded me to Brenda's concerns. The reality was Brenda had good reasons to believe both unexpected and unanticipated troubles would be part of our journey with implant surgery. Unfortunately for the both of us, it turned out that Brenda's worst fears would be realized before, during, and after surgery.

At this point some of you might be thinking, "Maybe an implant worked for him, but I won't consider that as an option for me." There are several unhealthy reasons why some men won't even consider the option of a penile implant. The first is related to posttraumatic stress disorder (PTSD). Many men do not realize they are coping with PTSD, which fuels their decision to avoid hospitals, surgeons, and every form of surgery. Those men will say they won't go within a mile of a hospital or a surgeon ever again, and they mean it. Without realizing it, PTSD is limiting their ability to see any positive outcome from a surgical procedure.

A second unhealthy reason why men won't seek help and find a possible cure for their ED has to do with two powerful emotions: embarrassment and shame. Some men feel so ashamed by their impotence that they avoid discussing it with anyone. Going to a doctor and admitting they have ED is unthinkable. Such men prefer to suffer in silence, possibly for decades rather than talk about the issue with anyone, including their partner.

A third reason is tens of thousands of men have had a bad experience with a previous surgery. This attitude is prevalent amongst men like me who were told to expect to regain their erectile function after bilateral nerve-sparing prostate surgery. When the expected outcome from surgery didn't materialize, they felt misled and/or betrayed. Unfortunately, these feelings become over-generalized, which means they are applied to every surgeon and surgical procedures on the planet. Men conclude that all surgeons are liars who are only out for a buck and do more harm than good. Men in this group also declare they won't go within a mile of another surgeon even if their life depends on it.

These are just a few of the many reasons why men will refuse to consult with their physician or a surgeon about their struggle with ED. This is a tragedy. Sadly, too many men and couples are wounded sexually, emotionally, and relationally due to ED when options exist to treat the condition. A question I ask every man coping with ED is this: Would you prefer to live in anger, bitterness, despair, isolation, broken relationships, and untreated impotence, or would you like to find a way to treat your ED? If you want to treat your ED, your first step involves consulting your primary care doctor about your erectile dysfunction. Finding the cause of your erectile dysfunction is the first step toward finding a successful treatment. If your ED is caused organically, meaning by an injury, illness, or a disease, implant surgery might be the only treatment option left.

I was in the group of men who didn't get the expected outcome from prostate surgery. Thankfully, I didn't make the mistake of over-generalizing. I knew that every surgeon wasn't a liar, in it for the money, or out to harm me. The option for penile implant surgery instilled within me the hope that I wouldn't spend the rest of my life impotent. I was confident that implant surgery would restore my erectile function. Therefore, I was surprised that I was adamantly opposed to consulting with the surgeon who had removed my prostate, even though he was an outstanding and highly-skilled surgeon. It was important for me to see a surgeon who wasn't associated with the surgery that had resulted in my impotence. If you are suffering from PTSD or over-generalizing, don't let your fears determine whether or not an implant is an option for you. Base your decision on the facts. If you can't do that, get help. It's far better to get help than to live the rest of your life lonely, depressed, or in despair, because it is completely unnecessary.

My impotence had been caused by prostate surgery. Even though both of my nerve bundles had been spared, they didn't heal in a way that restored my erectile function. Once I decided to have implant surgery, I needed to believe it would be a simple and easy process. I was convinced I was on a journey that involved a happily-ever-after ending. Once our journey began, I discovered that happily-ever-after endings only occur in fairytales. Real life is messy. I had to do a significant amount of work to heal the four years' worth of individual and relational wounds brought about by the ways in which I coped with ED. I was unpleasantly surprised to discover our physical, emotional, relational, and sexual healing would not occur automatically once I regained my erectile abilities.

Questions to consider:

1. At this point in your journey with ED, do you know what is causing the issue for you?
2. If you don't know, will you make a commitment to see a physician? If not, why?
3. If you've experienced multiple failures, how has that effected your hope that anything can change?
4. Is impotence something you want to live with for the rest of your life? If so, what can you do to maintain an enjoyable sex life?
5. If you have a sexual partner, how does your partner feel about the impact of ED on your relationship?
6. Make a list of the pros and cons of penile implant surgery. Ask your partner to do the same. Discuss your lists with each other.
7. What did you learn by sharing each other's lists?

CHAPTER 8

How to Choose Your Surgeon

Once a medical determination is made that an implant is an appropriate treatment, your doctor might provide you with a referral to a surgeon. If you prefer, and your medical insurance company allows it, you can do what I did, which was to go directly to a surgeon. My situation was unique in that my surgeon was also the urologist involved with my penile rehabilitation. He was familiar with my medical history and multiple treatment failures.

Once you find a surgeon and you are ready for your first appointment, understand this:

1. It's your surgeon's job to determine whether or not implant surgery is appropriate for you.
2. It's your job to determine whether or not you are comfortable with the surgeon's knowledge, skill, and experience with implant surgery.

There should be time in your appointment for you to ask questions. Here's a checklist of important questions:

- *How many implant surgeries do you perform in a year?* There is no magic number, but you will get a sense of whether the surgeon is actively involved with implant surgery or seldom performs such procedures.

- *What is your rate of infection?* The answer to this question should be less than two percent. The answer to this question should be a deal-breaker. If the surgeon with whom you consult has a high rate of infection, greater than five percent, I recommend you consult with another surgeon.
- *How often do your patients experience post-surgery complications?* The expected answer should be rarely. A skilled penile implant surgeon should not have frequent post-surgery complications.
- *What's the most frequent complaint you receive from your patients?* The complaint a penile implant surgeon should hear most frequently should revolve around the issue of penile shrinkage after implant surgery.

Finding a surgeon who performed penile implants was not as easy as I thought it would be. I began my journey by calling my local urologist. (Receiving a referral from your local urologist is a great way to find a surgeon experienced with penile implants.) I was unpleasantly surprised and disappointed to learn he did not perform implant surgery, but he was willing to refer me to a highly-experienced surgeon whose practice was at UCSF Medical Center.

Once I realized the option to have my surgery performed locally was closed, I wanted to consult with my rehab urologist from UCSF Medical Center. I called his office and asked if he performed penile implants. When they told me he did, I set up an appointment to see him.

You might wonder why I'd go back to my penile rehab physician when he had failed to tell me about penile implants on the day he told me I'd be impotent for the rest of my life. There are many reasons. First, he was familiar with my medical history. Second, I had total confidence in his training and surgical skills. When I'm looking for a long-term relationship with a physician, bedside manner is important. When I'm looking for a surgeon, I don't care about bedside manner in the least. I want a surgeon who has the training and skills to perform the surgery I'm about to undertake. Since he was familiar with my medical history, it took him less than a minute to determine I was a good candidate for penile implant surgery, particularly, a three-piece inflatable implant. I had heard from a few men who said their surgeons ruled out a three-piece implant because of their hernia mesh. I've had two hernia surgeries, which left me with wire mesh on both my right and left side. My surgeon assured me there would be no problem placing the reservoir in my abdomen. That's one of the many benefits of choosing an experienced and skilled surgeon.

We also discussed his infection rate, which was below two percent. He warned me that the biggest complaint he receives after surgery is from the men who are dissatisfied with either the length or girth of their erection with the implant. I waved that concern off, since I couldn't care less about the loss of size if surgery would allow me to attain a useable erection.

I left his office confident a penile implant was the right option for me and that he was the right surgeon. Once his office received the pre-authorization from my insurance company, a date would be set for my surgery. Most insurance companies will cover the cost of an implant if your erectile dysfunction is a result of an illness, injury, or disease.

There are other risks with penile implant surgery you should be aware of whether or not you choose to discuss them with your urologist. Here are some of them:

- Scar tissue formation
- Erosion occurs if the implant wears away the skin inside the penis. On rare occasions, the implant breaks through the skin. My surgeon told me it was much easier to fix if I caught the erosion while it appeared as a rash rather than after it broke through the skin.
- Pump or reservoir displacement: The pump or reservoir can move to the wrong place in your scrotum after surgery. Currently, my pump is not facing from front to back. It's on a forty-five-degree angle. I'm consulting with my surgeon as I write this to find out if there is a non-surgical solution to this issue. He thought the pump would reposition itself over time. That has not happened.

- Mechanical failure: It's possible the pump will not function properly.
- Uncontrolled bleeding after surgery: This condition might require additional surgery.
- Infection: This is a dreaded possibility because of the need for revision surgery to take the implant out, requiring a third surgery to put in a new implant.
- You might be responsible to pay a larger amount than you expected for your share of cost for your surgery.

Since the passing of the Affordable Care Act, it is important to find an in-network provider and hospital. Prior to the Affordable Care Act, my deductible was $2,500. After the law went into effect, my deductible was raised to $4,500. That's a lot of money out of my pocket. A high deductible might put the cost of surgery out of reach. The good news is, most hospitals are willing to accept a payment plan for your deductible. You can discuss the possibility of making monthly payments to the hospital before your surgery. I arranged to make a monthly payment to UCSF rather than come up with the entire amount up front. If you make the mistake of going out of network, your out-of-pocket cost could reach $10,000 or more.

Unless you have unlimited resources, you'll want to be certain your hospital and surgeon are both in-network. You'll also want to know what your deductible and out-of-pocket costs will be to prior to your surgery. A number of men on Medicare have told me they had their surgery without any out-of-pocket expenses. It's up to you to find out where you'll land with your out-of-pocket expenses for surgery.

If you have an existing health savings account (HSA), you can use that money to pay for your out-of-pocket expenses. If you are facing a substantial out-of-pocket cost and you do not have an HSA account, find out whether you qualify to open one. If you qualify, you can transfer your tax-deductible funds into your HSA. This enables you to pay for your medical expenses with money that is not taxed. Check the IRS regulations or contact an accountant if you are not familiar with health savings accounts. My entire out-of-pocket expenses were paid with funds from my HSA.

Once I chose my surgeon and I had the financial aspects of surgery covered, I thought the next step, waiting for surgery, would be the easiest phase of the journey. I had no idea of the emotional storm that was about to be unleashed inside of me and Brenda and how it would spill over into our marriage.

If you'd like additional help choosing a surgeon, the following website lists penile implant surgeons in one thousand cities across the United States: http://www.medicinenet.com/penile_implants/city.htm

Questions to consider:

1. Have you completed your health exam so you are certain there is a medical cause of your erectile dysfunction?
2. Are both you and your partner in agreement that a penile implant is the right option for you to take to restore your sexual relationship?
3. Are you certain both the hospital and the surgeon of your choice are an in-network providers?
4. Do either of you need to discuss any relational issues before you have surgery? (Some suggestions: Discuss what ED has taken away from your relationship and how you feel about those changes, what you hope an implant will do, and whether or not you need professional counseling to deal with the current state of your relationship.)

CHAPTER 9

Three Implant Options

Once you've decided to get a penile implant, you'll need to decide the type of implant you prefer. I wanted a Bluetooth implant with "an app for that." I imagined myself working in my office and then suddenly finding myself getting erect. At that point, I would know my wife was thinking about me and using her phone to tell me it was time to come home. When I shared this fantasy with a friend he said one word that killed my romantic fantasy: "hackers." Someday, a Bluetooth implant with "an app for that" will exist, but not today. Here are your current options:

The Semi-rigid Rod

With this type of implant, rods are placed within the erection chambers of the penis. As a result, you'll always be firm. If you think you'll have difficulty squeezing a pump, this is the right type of implant for you.

Advantages:

- It's easy to use: You simply bend it up for an erection and then bend it down when not in use. If you have arthritis or difficulty using your hands, this is the easiest option.

- If you don't have insurance and you are paying privately, this is the least expensive option.
- It has fewer mechanical parts, so breakdowns are rare.

Disadvantages

- You'll be erect at all times.
- Over the course of time, the rods might cause pain or erode through the skin.
- Your erection might be difficult to conceal.
- The size of your erection is limited to the size and rigidity of the prosthesis.

Ten months after my implant surgery, due to a cause unrelated to my implant, I developed a severe and sudden case of of carpel tunnel syndrome. I learned from experience that such medical conditions can make it impossible to inflate your implant. That doesn't mean you are forced into the option of a semi-rigid rod. If you can't inflate an implant by yourself but you have a partner who can, I recommend you skip this option and go for one of the inflatable implants. It is very uncomfortable to be erect 24/7. I've heard from many men who regret choosing this option. If it's the only option available to you, I suggest you talk to a number of men living with a semi-rigid rod before you make this choice.

The Two-piece Inflatable Penile Implant

This implant consists of inflatable cylinders inside the shaft of the penis and a combined fluid reservoir and pump placed in the scrotum.

Advantages

- Erections are larger and more natural than the semi-rigid rod.
- You can deflate your erection, thereby making this option more comfortable.
- It is easier to use than the three-piece inflatable implant.
- It does not require an abdominal incision.

Disadvantages

- Erections are not as firm compared to a three-piece device.
- The implant does not deflate as fully, as does a three-piece implant.

Compared to the semi-rigid rod, more men are satisfied with the two-piece inflatable implant. This is why I suggest choosing an inflatable option.

The Three-piece Inflatable Penile Implant

This implant consists of inflatable cylinders inside the shaft of your penis, a fluid reservoir under your abdominal wall, and a pump inside your scrotum.

Advantages

- It's easy to use. You pump it up for an erection and press the release valve above the pump to deflate.
- The larger, softer pump makes is easier to inflate than a two-piece device.
- It provides the most rigid erections compared to a two-piece device.
- It puts the least amount of pressure on penile flesh when not in use, making it the most comfortable option of all the implants.

Disadvantages:

- Requires more manual dexterity than other implants
- The most expensive implant and the most extensive implant surgery
- Higher risk of mechanical failure

I knew only one thing when I consulted my urologist about an implant: I didn't want the semi-rigid rod. There was no way I wanted to live with an erection all day, every day.

As I mentioned, since I have wire mesh on both my right and left side due to hernia surgery, I thought I the two-piece implant was my only option.

Sometimes (but not often) I'm delighted to discover I'm wrong. After a brief physical, my surgeon told me there would be no problem for me to have a three-piece implant. That was great news.

Once you've chosen the type of implant you want, the next step is to choose the brand. The two main penile implant manufacturers are American Medical Systems and Coloplast Titian. According to studies, both brands are similar in terms of the long-term functioning of the implant and patient satisfaction rates. What's the best way to decide? I have three suggestions.

1. Go online and do your own research on the brands available and make your decision based on your research.
2. Have a pre-surgical discussion of the brands and models with your surgeon. It's possible your surgeon will tell you he or she won't know which is the best model for you until they are in the process of surgery.
3. Ask your surgeon which model he or she uses most frequently and why.

I'm positive that both manufacturers have solid reasons to proclaim why their brand is best. I can't say whether it was due to laziness or because I trusted my surgeon, but I did not want to be the one to decide what brand was best. I was confident both brands would work effectively. Therefore, I decided to leave that decision with my surgeon.

Questions to consider:

1. How will you decide which implant is best for you?
2. Will you make up your mind before you see a surgeon, or will you allow your surgeon to help you decide which type of implant to choose?
3. Do you want to decide which brand of implant to use, or do you want your surgeon to help you with that decision?
4. What are your thoughts and expectations about penile implant surgery?

CHAPTER 10

PTSD and Me—and Brenda, Too!

I had an overly simplified and unrealistic expectation about my surgery. I thought I would move easily from being unhappy about being impotent to feeling happy once I had the implant. As I mentioned, I pictured the surgery as providing Brenda and me with a happily-ever-after ending. My first clue that my assumption was both naïve and inaccurate came in the form of sleeplessness. Once I had a surgery date, my anxiety level rose so high I couldn't fall asleep at night. I would go to bed at my normal time and lie awake tossing and turning. After thirty minutes of sleeplessness, I'd leave my bed and stay up until four or five in the morning. I'd get two to three hours of sleep and wake up by seven both tired and irritable. After a few days passed with me averaging less than four hours of sleep a night, I began functioning very poorly at home, at work, and with life in general.

The second clue that my assumption that surgery would result in an immediate improvement in my relationship with Brenda had to do with the increased number of fights we experienced. Every weekend, without fail, we spent a day or two fighting and angry with each other. We were unhappy with each other and our marriage.

Brenda thought both of us had been suffering from PTSD for years. I didn't believe her. When Brenda and I received our professional training, a diagnosis of PTSD was limited to veterans and those who had suffered major trauma. According to an article called "Cancer Treatment Leaves Survivors with PTSD Scars,[1]" before 1994, cancer patients were specifically excluded from the psychiatric definition of PTSD. At that time, so few survived their treatments that there was rarely a "post" phase to deal with. Today, there are fourteen million cancer survivors in the United States, and the diagnostic criteria for PTSD has been expanded to include them, their partners, and family members involved in caretaking.

After spending a few sleepless nights reliving my experiences with prostate surgery at UCSF Medical Center, I decided to research the topic of PTSD among cancer survivors. When I Googled the topic, I got 218,000 hits. That was my first clue that PTSD among cancer survivors was a reality. A Reuters article stated, "More than a decade after being told they had the disease, nearly four out of ten cancer survivors said they were still plagued by symptoms of post-traumatic stress disorder, or PTSD."[2]

According to the National Institutes of Health (NIH), three different types of post-traumatic stress disorder exist. If symptoms last less than three months, the condition is considered acute PTSD. If symptoms last at least three months, the disorder is referred to as chronic PTSD. If symptoms manifest at least six months following a traumatic event, the disorder is classified delayed-onset PTSD.[3]

According to the Mayo Clinic, the most common events leading to the development of PTSD include:

- Combat exposure
- Childhood neglect and physical abuse
- Sexual assault
- Physical attack
- Being threatened with a weapon
- A life-threatening medical diagnosis

"Many other traumatic events also can lead to PTSD, such as fire, natural disaster, mugging, robbery, car accident, plane crash, torture, kidnapping, terrorist attack, and other extreme or life-threatening events."[4]

I have come to believe that ED can trigger PTSD. Here's a list of symptoms that are characteristic of PTSD from the Department of Psychiatry Penn Behavioral Health.[5]

- Intrusion or re-experiencing: Recurrent recollections of the event (referred to as *flashbacks*)
- Dreams, intrusive memories, and discernable prolonged distress and physical reactions to cues that resemble the traumatic event
- Avoidance: fear and avoidance behavior
- Avoidance of people, places, thoughts, feelings, or activities associated with the traumatic event
- Changes in mood and cognition: negative alterations in emotions or thoughts
- Exaggerated negative beliefs and self-blame for the traumatic event, detachment from others, loss of interest, persistent negative emotional state, reduced ability to feel positive emotions
- Arousal and hyper-reactivity: agitation, state of constant wakefulness and alertness
- Hypervigilance, being easily startled, acting irritable or aggressive, recklessness
- Sleep disturbances, difficulty concentrating

When our therapist confirmed we were suffering from PTSD, I was the only one who was surprised, though neither of us were surprised to learn that our trip to San Francisco to see my implant surgeon was a PTSD trigger. Over the last four years, all of our trips to San Francisco had been prostate cancer-related. Now I was returning to "the scene of the crime," as it were. I was going back to UCSF Medical Center for another surgery. My last surgery at UCSF brought about very difficult challenges to me physically and emotionally. Brenda had also suffered from emotional, physical, and relational abandonment. Our sex life and our marriage had suffered greatly, and now we were returning to UCSF for another surgery, hoping for a totally different outcome. To say the event brought back some old, unpleasant, and painful feelings would be the understatement of the year.

After prostate surgery, I had no desire to return to San Francisco. I thought the only reason I had made that decision was because of my dislike of traffic and big cities. I had no idea that my decision to avoid San Francisco was PTSD driven until I had to drive back to San Francisco for both consultations and implant surgery. My anxiety before each trip was sky high.

Brenda and I learned the hard way that PTSD can have a negative impact our thinking and our mood. It made no sense to me that we were on the verge of ending our experience with impotence, and yet our fighting and marital unhappiness had risen to relationship-threatening levels. I'm glad we sought out professional help. We had neither the skills nor the resources to resolve the crisis on our own.

Based on dozens of conversations with men who are coping with permanent erectile dysfunction, I have come to believe PTSD plays a significant and unrecognized role in men's resistance to consider penile implant surgery. Frequently, I hear comments like "I won't go within a mile of another surgeon" or "I'll never let a surgeon touch me again." The thrust of these comments has to do with the anger and avoidance of the misery brought about by prostate surgery. These men want to avoid any further negative consequences or damage as a result of a second surgery. Some men no longer believe anything positive can come from a surgery. Therefore, a fairly large group of men would rather live in misery than relive or re-experience their disappointment and anger over the loss of their manhood as a result of having their prostate removed, even if it means not undergoing implant surgery, which can restore what was lost.

I had the opposite reaction. Based on my research, I was certain I'd have a positive result from implant surgery. That's why I was taken by surprise as I experienced overwhelming levels of anxiety and fear. It didn't make any sense that I'd be coping with PTSD when I was anticipating a restorative procedure. I don't how many times I have relearned this lesson. Our emotional life isn't governed by logic. I mistakenly thought the faster I got into surgery, the more rapidly our journey with PTSD would end. I didn't get to test my theory, but my therapist assured me it was wrong.

A few weeks before surgery, I received a phone call from my surgeon's office. He had been involved in a bicycle accident, which required hospitalization. My surgery was postponed for a minimum of six weeks. That was the worst possible news I could have received. I didn't think I would survive another six weeks of sleep deprivation and anxiety. I had to do something.

I called my physician and asked him to prescribe a two-month supply of medication to help me sleep. Do not underestimate the value of a good night's sleep! However, once again, I developed an unrealistic hope that I could make the host of symptoms related to PTSD go away with a few good nights of sleep.

In therapy, we learned we had some unresolved relational and personal issues we had neglected to deal with. In addition to coping with PTSD, we had a lot of baggage we needed to resolve about the ways in which my response to ED had hurt Brenda. We resolved those issues in counseling. Additionally, we worked to create new and positive associations with our journeys to San Francisco. We began our drives into the city listening to music other than the CDs we had listened to on previous journeys. In addition, we made sure to include a fun activity with each trip.

For example, on our trip to San Francisco for implant activation, we reserved a hotel for our first experience with the implant. Instead of dreading our trip to the city, both Brenda and I looked forward to a romantic night together. We weren't disappointed! Our trip for implant activation and our romantic getaway was my favorite trip to San Francisco. By the time we left the city the following day, we had broken the power of PTSD over our trips to San Francisco.

If you believe you are experiencing PTSD, the following websites offer tips for self-help and a list of resources for professional help.

- HelpGuide.org: HYPERLINK "http://www.helpguide.org/articles/ptsd-trauma/post-traumatic-stress-disorder.htm"
- The National Center For PTSD: HYPERLINK "http://www.ptsd.va.gov/public/treatment/therapy-med/treatment-ptsd.asp"

PTSD does not go away by itself. If you are experiencing symptoms of PTSD, contact a professional who has experience treating the issue. The professional help Brenda and I received helped us as individuals and in our marriage relationship.

Questions to consider:

1. What stressful/chronic medical issues resulted in you experiencing ED?
2. Based on the information contained in this chapter, is there a possibility that PTSD is part of you and/or your partner's life?
3. If you suspect that either or both of you are suffering from PTSD, how do you plan to deal with this difficult and stressful issue?
4. If you need more information about PTSD, what is preventing you from checking out the online resources at the end of this chapter? (If you don't have internet access in your home, go to a public library.)

CHAPTER 11

Check Your Unrealistic Expectations at the Door

Some men go through with surgery only to discover that they rarely, if ever, use their implant. It's important to understand what implant surgery will and will not do. For example, your newfound abilities won't magically increase your desire or your partner's desire for sex. If your libido was low prior to surgery, the ability to achieve an erection via the implant will do nothing to increase your desire for sex. I know this to be true from personal experience. From the time I was a teenager until I reached my mid-fifties, I had a high interest in sex. After I was diagnosed with prostate cancer, my libido dropped to zero. Four years later, it had not returned. My implant had no effect on my desire for sex. Even today, my sexual desire remains at the lowest level I've experienced in my lifetime.

In the months following implant surgery, I learned an important lesson. You can thoroughly enjoy sex without having a strong desire to do so. At age sixty-four, even with little desire, I'm having sex more frequently than at any other time in my life. Additionally, I'm enjoying every encounter. No one is more surprised than me to discover that I'm experiencing greater enjoyment and pleasure now, when my desire is close to zero, than I did at any time when my desire was at its peak.

Based on my experience, I've concluded that many couples make a serious error if they decide to give up on sex because one or both of them have lost their desire. There's lots of pleasure to be found in the physical act of two becoming one. Experiencing an orgasm even though you don't had a strong desire to have one is also highly pleasurable. I discovered another important lesson as well. The act of joining together physically can be highly pleasurable with or without an orgasm.

One important issue to resolve if your partner is in menopause involves doing what's necessary to ensure that sex is not painful in any way. Brenda and I have found that using a lubricant is a very important component of each and every sexual encounter. I make sure I'm well-lubricated so that penetration feels pleasurable rather than painful.

Most men coping with ED spend a significant period of time ignoring their partner's emotional and physical needs as part of the unhealthy ways of coping with erectile dysfunction. You're making a huge mistake if you assume that your partner will welcome your sudden interest in resuming your sexual relationship without a discussion of any relational issues that need to be resolved.

Some couples have the capacity to do this on their own. Other couples, like Brenda and I, need professional help. Do what's necessary to restore both the emotional and physical harmony in your relationship before *and* after you have your surgery. Implant surgery is not a magic bullet that will turn back the clock and restore your relationship to a happier, more fulfilling state. To accomplish that task, you need to do the work required to bring about relational healing. If you skip this task, it's likely you'll find yourself in the group of men who rarely use their implant.

If you want to begin repairing the relational damage brought about by the hurtful ways you coped with ED, ask your partner the following questions. Since ED visited our relationship:

1. How have I met your emotional needs?
2. How have I met your relational needs?
3. How have I met your spiritual needs?

4. How have I met your sexual needs? (This includes touching, kissing, and all other forms of physical affection, including intercourse.)
5. Are there any unresolved issues, such as feeling abandoned, angry or resentful, through which we need to work?

As you discuss these issues, the presence of avoidance, out-of-control anger, defensiveness, hostility, bitterness, or blame are all indicators you probably need outside help.

I experienced some meaningful, positive changes within the first month after my implant activation. Regaining my erectile abilities increased my and Brenda's enjoyment of sex dramatically. In addition, I regained my confidence in the bedroom, which led to me feeling a lot better about myself as a man and as a sexual partner. I believe you'll experience some immediate positive changes as well, should you choose to go through with implant surgery.

Apart from the relational issues, other issues can also lead some men into despair after implant surgery. Many men who have coped with ED for years discover that, after implant surgery, their erection is significantly smaller than they expected. This is a result of penile shrinkage. Men who place a high value on the size of their erection find themselves moving from one source of depression, ED, to a another source of depression, the smaller size of their erection. Rather than feel grateful they can enjoy sex with an erection, they remain angry, bitter, depressed, and/or obsessed with their smaller size and girth.

If the size of your post-surgery erection is important to you, make sure you discuss and understand where you will land with both your penile length and girth after your implant. If penile lengthening is important to you, you'll need a surgeon who has experience with both the implant and penile lengthening. I must confess I can't relate to men who experience post-implant depression. However, I certainly understand why men are disappointed about the size of their erection with an implant. I was disappointed when I discovered how small my erection was. As I looked down at my erect penis, I felt as though I were ten years old.

However, once I discovered my smaller size made no difference to Brenda or me in regard to the pleasure we experienced, my disappointment went away. What was most important to me was the fact I had regained what was lost: the opportunity to enjoy becoming one physically with my wife. From my perspective, it's a total waste of time and energy to focus on the reduced size of your erection.

Imagine living in a darkened jail cell for a decade. Upon your release from prison, you discover the sun doesn't shine as brightly as you remembered. Rather than enjoy your newfound freedom from prison, you invest all of your time and energy on your disappointment with the sun. That's my sense of what a man does if he loses his joy over the success of implant surgery by becoming depressed about the smaller size of his erection.

I chose to experience daily gratitude that I live in an era and a country where penile implants are possible. Rather than live with ED for the rest of my life, I've been granted a reprieve. For me, ED was like being sentenced to prison for the rest of my life. The implant was my executive pardon. Now I'm free! I feel grateful for this freedom every day of my life, whether or not I put my implant to use.

After a few months of enjoying the act of two becoming one, my experiences are consistent with the research on patient satisfaction with the implant. If you Google the topic "satisfaction rates with penile implants," you'll find a variety of studies that place user satisfaction rates between eight-five to ninety-seven percent. Inflatable implants have the highest satisfaction rates.[1] Partner satisfaction rates are high as well, ranging from the mid-seventy to the mid-ninety percent range.[2] In other words, the inflatable implant has both the highest user and partner satisfaction rates than other surgical methods of treating erectile dysfunction.

Most men who go through implant surgery have tried and failed with every other attempt to treat their ED. After you experience multiple disappointments and failures, it's logical to conclude the best thing to do is to give up. The idea of surgery to cure ED is a frightening possibility. Your experience with multiple failures makes it difficult to believe surgery will bring about a successful outcome. The idea that surgery will cure ED is especially difficult to accept for men who developed ED as a result of surgery. I've come to believe that's the reason why so many men who have had surgery to remove their prostate won't consider another surgery to regain their erectile function. The reality is implant surgery is an amazingly successful way to treat ED.

The decision to have implant surgery was one of the best medical decisions I've ever made. Both Brenda and I are delighted with the results. Even though we chose a highly-skilled surgeon, I didn't have any easy recovery. In the next chapter, I'm going to share my experiences with an emphasis on everything that went wrong. I believe you'll learn some valuable lessons and avoid unnecessary complications, pain, and grief by reading about my post-surgical experiences. But first, Brenda will give her perspective.

Questions to consider:

1. Discuss what you and your partner expect to happen in your relationship once you regain your erectile abilities.
2. How frequently do you expect to use the implant following surgery?
3. How do you and your spouse think the implant will change your relationship? How do you feel about those changes?
4. Discuss your expectations regarding the frequency of sexual activity following implant surgery.
5. Do either of you have any concerns, fears, or worries you need to discuss with each other?
6. Are you hiding any fears, worries, or concerns from your partner?

CHAPTER 12

Before Implant Surgery: A Wife's Perspective

I've been blessed with a few close friends, but in the five years that I (Brenda) have been the wife of a cancer survivor, I have never discussed Rick's impotence with anyone. Whenever people asked "How is Rick doing?" I'd answer with a short explanation of his mental state and his physical progress. I'd never discuss our "new level" of intimacy or lack thereof. Certainly, it was natural to feel inhibited discussing the changes brought about in our sex life as a result of cancer and prostate surgery.

It's ironic that I grew up in the era of love, sex, rock 'n' roll, and proclaimed openness, yet when push came to shove when facing the issue of impotence, I found myself alone and isolated. All I heard was the sound of silence. The silence from friends was tolerable, but the silence from Rick, who refused to discuss his impotence, was hurtful and harmful to our relationship.

In the first two years after Rick's prostate surgery, it took a fervent effort on my part to keep Rick involved with penile rehabilitation. He was discouraged easily and often ready to quit. I approached Rick every day with a playful spirit to continue rehab to prevent a venous leak, which is a frequent cause of impotence with men who have had their prostate removed.

Our first attempt to restore our sex life after Rick's prostate surgery was the vacuum pump. Rick hated the device and never attempted to have intercourse using the pump. He tossed it in the garbage after a few short weeks. We tried oral medication with some success but mostly failure. After that, Rick tried penile injections. For three months, it worked, and our sex life was restored. Then the injections were no longer effective. We tried another medication, but it was so painful that we gave up on penile injections. We went back to oral medication, trying different brands in different daily doses. Then Rick tried testosterone. There was no change in his desire for sex or his erectile function.

With each form of treatment, we went through high hopes of success followed by deep valleys of failure. Not only that, Rick did not like me approaching him sexually. As a result of his fear of failure and loss of confidence in his ability in the bedroom, he did and said everything he could think of to push me away. It was a lonely and isolated time for me both physically and emotionally.

The third year after surgery was the hardest year for me and for us as a couple. I lost my best friend, with whom I used to share everything in my heart. The icing on the cake was when my brother-in-law, who had been diagnosed with a brain tumor, began to decline rapidly. He was at the end of life and had a couple of months to live. This had a profound effect on Rick. He wanted to regroup as a couple, and we determined that no matter how stressful our situation was, we weren't in the end of life, so it was time to approach each other once again. In our fear, we held onto each other for a while. However, soon after my brother-in-law passed, we were back to our old patterns. I think that was when PTSD took root.

When I first noticed I was having trouble, I proclaimed jokingly that I thought I had PTSD. I noticed I was vigilant about everything around me. Feeling stressed all the time, I also noticed I did not relate to others like I used to with conversations flowing freely. Instead, I felt like I was missing chunks of the interactions. My desire to go out and be with friends also diminished.

I started looking online for connections between cancer and PTSD. To my astonishment, I found that a relationship existed between the two. Not only that, the articles spoke of cancer patients and their partners both developing PTSD. In many ways, I felt great relief, because I finally had a name for what I was experiencing.

In the beginning of the fourth year following Rick's prostate surgery, Rick had an appointment at UCSF with his penile rehab urologist. He was told his nerve bundles had healed completely and there would be no further improvement in his erectile function. Rick was going to be impotent for the rest of his life. We drove home in silence.

A few months later while I was visiting our son in Vermont, Rick called and told me that he wanted an implant. Rick had made his decision unilaterally and without any discussions with me. My first response was neither loving nor supportive. It sprang from my experience with PTSD. I told Rick there was no way I was going through another surgery with him! By that time, our relationship was fully deteriorated. He was still pushing me away, and I was not a very loving, kind person towards him.

His desire for surgery pushed me to an edge I had never reached before. Unbeknownst to either of us, Rick stepped directly on one of my PTSD triggers. It was like a land mine going off. I couldn't go through any part of a surgery again.

At that low point, I called our counselor. In our first session, he agreed with my assessment that both Rick and I were suffering from PTSD. Rick was still in favor of surgery, but I was done, so done. However, with help, I kept going. I prayed and knew God was still there, even though I couldn't feel His presence. The conflict in our marriage had reached a level that frightened both of us. Neither Rick no I had the ability to stop our destructive and unloving behavior toward each other.

Our counselor helped us to see the thoughts and behaviors that interfered with our relationship. He saw the harsh tone and helped us to correct it. He told me my need for Rick to be available when he was unable was another one of my triggers. He also explained that my desire to have Rick hear my heart once again was something that PTSD rendered Rick unable to do. Hearing that made me realize I was going to an empty vessel asking for something he could not give. I also understood that if our relationship was going to get better, it was neither my job nor my responsibility to change Rick's reactions to coping with erectile dysfunction. However, it was within my power and responsibility to change my reactions. At that moment, I saw that cancer had changed both of us. Eventually, I agreed with Rick's plan to have penile implant surgery.

As the weeks before surgery approached, Rick became highly anxious. He was unable to sleep. His lack of sleep might have contributed to his highly irritable mood. He had to see his primary physician to get medication to help him sleep through the night.

The anxiety, fighting, and negativity, all of which were symptoms of us coping with PTSD, made the weeks before surgery difficult, but I was determined I would not let the negativity win. Our counselor led me through some stress-reduction tips to get through being back in the same hospital for the implant surgery where Rick's prostatectomy had taken place. I set up the various scenes to be able to cope on some level through another surgery with Rick. We were told implant surgery was a relatively simple procedure that involved little or no post-surgical pain. I believed those reassurances and felt quite prepared as we went to the hospital.

As I expected and feared, Rick's surgery was neither simple nor pain-free. All hell broke loose in the hospital, on the ride home, and in the days following the surgery. In the course of our marriage, Rick has had at least seven surgeries. Most have resulted in unexpected complications. Rick is also very stubborn. He rarely follows his post-operative restrictions. His lack of compliance usually adds additional complications to his post-surgical recovery. As a nurse, Rick is the worst patient I've ever had to care for. Going through one more surgery with multiple complications was overwhelming to me.

One step at a time was all I could face. After many struggles, we finally made it through the surgery and recovery process. Both of us were joyfully anticipating a new sex life together.

Questions to consider:

1. What, if any, fears are inhibiting you from going forward with implant surgery?
2. Do you anticipate any losses in your sex life with a "bionic" penis?
3. Are you concerned your partner might not want you to have this surgery due to the lack of sexual desire or concerns regarding painful sex?
4. How has impotence affected your ability to be affectionate with each other?
5. Has impotence brought a new tenderness and understanding or negativity and fighting into your relationship?
6. What concerns you regarding the way an implant will change your relationship?
7. What positive changes might happen if your sex life together is restored with an implant?

CHAPTER 13

Implant Surgery, Here I Come

A few weeks prior to my surgery, I (Rick) attended a worship service where a man facing two surgeries to treat cancer asked the church body for corporate prayer. In the life of our congregation, it was an important and moving time of song and prayer. I was weeks away from facing my own surgery. I was conflicted as to whether or not I should ask the congregation to pray for me.

The spiritually mature part of me knew there were specific issues we could pray about. For example, praying for the surgeon, praying there would be no complications, and praying the surgery would be successful. All of these issues were appropriate for corporate prayer, but my embarrassment and desire for privacy got in the way. I couldn't see myself asking everyone to pray for the success of my penile implant surgery.

Both Brenda and I were grateful when our pastor asked if he could visit with us the day before my surgery. He came to our house and prayed with us. It was a wonderful and comforting time. Before he left, I asked him not to announce my surgery or to ask for prayer during the weeks I'd be recovering at home. I wanted to give as few explanations as possible about the nature of my surgery.

When friends asked specifics about my surgery I explained, "I'm having a restorative surgery related to my prostate cancer." I made sure two words never appeared in my explanation: "penile" and "implant." Thankfully my vague explanation was enough. Only one person continued to press until it became necessary to explain I was going in for penile implant surgery. I was surprised at the follow-up question, which raised my level of embarrassment through the roof: "What's a penile implant?" That round of questions solidified my intention to protect my privacy regarding the specific nature of my surgery.

As with any surgery, risks are involved. As a way of coping, I tend to avoid taking a serious look at the risks, but one risk put fear into my heart: the risk of infection. I knew the implant had an antibacterial coating. I also knew I'd have a round of antibiotics after surgery, and I knew my surgeon's rate of infection was below two percent. Brenda and I also knew I had a tendency to land in places where odds were slim.

As Brenda and I went into treatment for PTSD, our therapist encouraged us to develop new and positive associations with our trips to San Francisco. In keeping with that goal, we decided to stay at a hotel the day before surgery. That served two purposes. The first was to give us a day to play and enjoy the city. The second had to do with convenience. I was scheduled to check in at 6:30 a.m. If we stayed at home, it would have been necessary to wake up at 3:30 a.m. and drive ninety miles to UCSF. Since we were only minutes from the hospital, we both appreciated the opportunity to sleep in a bit later.

When I woke up, I decided there was no reason to change into my street clothes. I knew one of the first things that would happen after check-in would be the issuance of a hospital gown for me to wear. It would be a lot easier for me to take off my pajamas than my street clothes, so I went into the hospital and waited in the check-in line in my pajamas. I was surprised to observe that I was the only person in line who had the common sense to stay in his pajamas. My daughter, Kate, thinks my decision to stay in my pajamas had nothing to do with practicality or wisdom. She is convinced I lack appropriate manners and good judgment. I have a long history of behaving in ways that totally embarrass her. Unfortunately for her, I take her embarrassment as a sign I'm doing the right thing. Therefore, since Kate was totally humiliated and embarrassed by this decision, I felt totally confident that going into the admissions office in my pajamas was the right thing to do.

For most men, penile implant surgery takes approximately forty-five minutes to an hour. The surgery is typically an outpatient procedure, performed using general anesthetic. Since we had chosen a facility more than ninety miles from home and I have a history of going into urinary retention after any surgery that required general anesthesia, my surgeon thought I should stay overnight.

I want to warn you in advance that my experiences were not typical, because they could scare some men away from implant surgery. Any surgery involves risks and the possibility that something might go wrong. I'm sharing my experiences in the hopes you can avoid some of the unnecessary and unpleasant issues we faced after my surgery.

I've had approximately nine surgeries in my lifetime. One of the most unpleasant effects is that I tend to vomit in the recovery room. When I shared this issue with my anesthesiologist, he said he'd do two things to prevent it. The first was to administer the anesthesia via a needle rather than a mask. Second, he'd give me anti-nausea medication. Following my surgery, I was relieved and delighted that and I did not throw up in the recovery room.

To avoid a painful episode of urinary retention, I asked for s catheter to be inserted during the surgery.

When I was brought back to my room, I didn't realize I had not been set up with a call button. When I wanted my first dose of pain medication, I had no way to call the nurse. Thankfully, I had my cell phone placed on my bed after surgery, so I called UCSF and asked for the surgical floor. When a nurse answered, I told her I was a patient on the floor and needed both my pain medication and a call button. Within minutes, a nurse came to my room. I was delighted by the way in which I resolved that problem and how quickly help arrived. Between my uneventful stay in the recovery room and how quickly my pain medication was provided, I was convinced I was going to experience an uneventful and peaceful twenty-four hours.

Soon after I was transferred from the recovery room to my hospital bed, my bladder began to spasm every few seconds. The spasms were so painful I couldn't help but cry out in pain as they occurred. As soon as I experienced my first spasm, I pressed the call button. When the nurse came in, I had already experienced multiple spasms. I don't know how they stopped them, but I do know it took the team approximately four hours to get them under control. Those were the most painful four hours I've ever experienced. After my spasms were under control I was convinced nothing else would or could go wrong. Once again, my assumption was incorrect.

Somewhere in the middle of the night, my nurse noticed my catheter bag was empty. She checked for any kinks or blocks in the line. None were found. She told me I needed to drink more fluids. For the next thirty minutes, I drank as much as possible. My catheter bag remained empty. They called my surgeon. They gave me additional fluids via IV and performed blood tests.

Shortly thereafter, they gave me a bolas of magnesium. After that, they walked me around the unit. Despite their efforts, I did not produce a single drop of urine. I developed my own theory as to why I had stopped producing urine. I was convinced I'd gone into renal failure. I wondered if my next stop was kidney dialysis and whether my kidneys would ever work again.

More than an hour passed, and I had not produced a single drop of urine. My fears kicked my imagination into full gear. I wondered if I'd live long enough to see Brenda in the morning.

Since I thought I might die and I had nothing to lose, I decided to take matters in to my own hands. Even though I have a history of going into urinary retention after surgery, I asked if they would remove my catheter and allow me to attempt to urinate on my own. They called my surgeon, who granted permission. When I went into the bathroom, I prayed that

I would be able to defy my medical history and urinate on my own. I was overjoyed when I brought back a container full of urine for them to measure. From that point on, my intake of fluids and output of urine was monitored carefully. To my relief, my medical crisis was over. By early afternoon, they decided I was ready to be discharged.

We had a ninety-mile drive home, so we took precautions to ensure I would be comfortable. They gave me narcotic pain medication right before my discharge. My scripts were called in to my local pharmacy. They were ready to be picked up when we arrived in Modesto.

What we didn't know was that a special event was taking place in San Francisco. Traffic was bottlenecked. We expected to arrive home two hours after we left the hospital. Instead, we were stuck in San Francisco traffic. My pain medication wore off before we got out of the city. I couldn't believe it. After all the planning for a comfortable drive home, I faced continuous and unrelenting pain for the next two hours. After a stressful hospital stay and a pain-filled journey home, I was beginning to doubt whether my decision to get a penile implant was a good idea.

Shortly after our arrival home, Brenda examined my surgical site. She was shocked at what she saw. During my four hours of bladder spasms in the hospital, I had leaked a significant amount of urine onto the surgical dressing. The gauze had turned into a paper mâché-like substance, which was stuck all over my scrotum. There was no way to lift it off. She needed to use warm water and a washcloth to scrape the gauze off. It was a slow and painful process.

I was relieved it was my wife rather than a hospital nurse who was washing and scraping gauze off my scrotum. Standing in the bathroom with her, I was on my thirty-sixth hour without sleep, and I was not looking forward to going to bed. I sleep on my stomach, never on my back. However, given the size of my scrotum, sleeping on my stomach was not an option.

Thankfully, I had been told sleep disturbance after surgery was highly likely, so I had asked for an additional two-week prescription to help me sleep. By the time I took my medication and went to bed, I had been awake for forty consecutive hours. Moments after my head hit the pillow, I was fast asleep. I didn't know it at the time, but that would be my last night of good sleep for a few weeks.

I have a medical history that includes seasons of back pain. During those seasons, I'm unable to walk without support. It wasn't an easy concession to make, but for me to make it up the stairs to sleep in my bedroom or to take walks with Brenda, I had to buy a cane. I was both pleased and proud that I had given in to reality and accepted the help I needed. Earlier, I had received a vivid reminder of what happens to men when they refuse to accept this reality.

At church one day after the worship service, a man who was walking in front of us lost his balance. He hit his head on the concrete so hard he was knocked unconscious. An ambulance was called. While we waited for the ambulance to arrive, his wife bemoaned the fact that he had a history of losing his balance. His wife was frustrated by his stubborn refusal to rely on a cane. As I saw him lying on the pavement, I realized the high price men pay in the service of their pride. I understood how difficult it was to accept the reality he needed help with a simple activity like walking. On that day, and every day thereafter when I had to use my cane, I was glad that I didn't need to hit my head on cement before accepting the reality I needed a cane to help me walk.

One day, however, I foolishly allowed a single comment from my neighbor to turn my victory into defeat. It was a sunny day, and I was experiencing back pain. Brenda wanted to go for a walk, so I took my trusty cane with me. As we walked together, I felt that familiar gratitude that using a cane made it possible for me to enjoy a walk with my wife. My feelings of gratitude were interrupted when a neighbor laughed at me and pointed at my cane. "Why are you using a cane?" Then he made a hurtful and life-changing comment: "You look like an eighty-year-old man!"

Words are powerful. They can encourage and motivate as well as shame and embarrass. I allowed my neighbor's thoughtless comment to wound me. I have a very attractive wife. I didn't want people to wonder, "How did that old man land such a beautiful woman?" After that walk, I threw away my cane and made up my mind that I would never use it again.

Thankfully, I remembered the lesson from church and what happens to men if they refuse to accept reality. There was no doubt in my mind that there would be many seasons in which I would need help walking, but I wanted to look cool rather than like an old man. I went online and purchased a stylish walking stick. In my seasons of back pain, it works as well as the cane did, with one important difference: When I'm out with my walking stick, I get frequent compliments. People who see me using it come up to me and ask, "Where did you get such an interesting walking stick?" No one points at me and laughs. No one tells me I look like an old man. Instead, I continue to receive all sorts of positive attention.

Upon my arrival home following surgery, my walking stick and I became inseparable. It was impossible for me to get up our stairs or get out of a chair without it. Any time I needed to walk anywhere, I had to use my walking stick for support.

During the first two weeks, my pain level remained at nine out of ten. From my perspective, oxycodone was useless. If post-surgical pain was like stereo music, I had expected I would continue to hear the music, but it would be in background, so there would be times when I'd forget the music was playing. Instead, for two weeks, it was as though someone had turned up the stereo full blast and I could hear nothing but loud, continuous music blasting in my ears every waking minute. I was in constant, unremitting pain that never went away in the daytime or at night.

When taking narcotic medication, it is illegal to drive. If it were legal for me to drive, it wouldn't have mattered. I was in way too much pain to get behind the wheel. Brenda had to drive me to San Francisco for my two-week post-surgery checkup. A short time after my surgery, my surgeon had moved his outpatient office. That was our first visit to his new location. When we parked in the garage, neither of us knew how far we were from his office. As we began the walk, I was in serious pain. I don't know how long it took to reach his office, but given how slowly I walked, I'm certain it took us more than twenty minutes.

By the time we reached the waiting room, I was exhausted and overwhelmed by pain. When the receptionist informed us the doctor was running late, I became irritable and agitated. I had spent so much time and effort to arrive at my appointment at the proper time, only to discover it hadn't been necessary. If I had known my surgeon was running late, I would have taken the time to stop, sit, and rest rather than push through the pain to reach his office on time. It would be both kind and considerate if medical office staff sent text messages to patients when doctors are running unusually late. I won't hold my breath waiting for that to happen.

By the time we were escorted to the examination room, I was furious. When I asked myself what was contributing to my fury, I remembered a comment he had made to me prior to surgery: "The majority of my patients find their post-operative pain so mild they don't need prescription medication to manage it." For the last two weeks, I had suffered unrelenting pain and sleep deprivation.

At his request, I dropped my pants so he could examine the surgical site. As he examined my swollen scrotum, I asked what had gone wrong. I appreciated his honest response. He said that during the surgery, he had not seen any external bleeding. Therefore, he thought it was unnecessary to insert a drain. Unfortunately for me, there was internal bleeding that he didn't see. All of that fluid had settled into my scrotum. He went on to say that given the level of swelling, it was highly unlikely I would be ready for activation in a month. He warned me it might take an additional two to three weeks.

When I did the math, I realized he was telling me I might be dealing with my painful and oversized scrotum for seven weeks. I worried whether I could cope with constant pain and sleep deprivation for another five weeks. Once again, I wondered if my decision to have an implant had been a serious mistake. It occurred to me that Brenda had warned me that I would have a rocky road before, during, and/or after surgery. My wife was gracious. With all that went wrong at the hospital and at home, she had many opportunities to say "I told you so," but she did not utter that phrase even once, and I was grateful. I hate it when I get it so wrong and Brenda gets it totally right. She tried to warn me to expect the unexpected, but I wouldn't listen. In my one-day stay at the hospital, I endured four hours of painful bladder spasms, stopped producing urine after surgery, had a painful ride home, and now my scrotum was swollen to the size of a grapefruit, putting me in constant pain. After the bladder spasms, the pain from my swollen scrotum won second place for misery, pain, and suffering.

It was obvious to my surgeon I was in extreme pain. I was grateful when he prescribed a month-long prescription of oxycodone. We filled the prescription before we walked to the parking lot. I downed one hoping it would make the walk and the drive home slightly more bearable. As we drove away from the appointment, I was discouraged that the surgeon's decision not to put in a drain had cost me so much unnecessary pain and discomfort. Not only that, I was also facing a two-week delay of my implant activation. My mood was dark for the entire drive home. Note: If you decide to have implant surgery, make sure you ask your surgeon about the possibility of a drain. The painless insertion of a drain during my surgery would have saved me three weeks of constant pain and suffering.

I've heard from men whose urologist wanted them to activate their implant during their two-week post-surgery visit. Some surgeons believe you can minimize penile shrinkage with an early activation. Every man who shared his experience with early activation stated it was intensely painful. After a single attempt, they postponed any further attempts to activate their implant prior to six weeks after surgery. I have read a few studies about early activation. They all end the same way. As a result of pain during early activation, the drop-out rate of men participating in the studies was too high. Pain and discomfort will prevent most men from attempting activation in the first two weeks after surgery.[1]

At some point in the third week after surgery, Brenda noticed a discharge from my surgical site. Since it was a small amount of pus, I mistakenly assumed it was a small problem. Brenda was insistent that pus, no matter how small, was evidence of an infection. She insisted that I call my surgeon's office immediately to get back on antibiotics. She knew a post-surgical infection was a potentially serious complication. I was put on antibiotics that day. We both prayed the treatment would work so I could avoid a salvage surgery to remove my implant and a third surgery for a new one. One implant surgery was more than enough for me.

Two important and wonderful things happened over the next few days. First, the infection was wiped out. Within the first few days on my second round of antibiotics, the discharge stopped. Brenda and I were relieved. One night, I went to bed with a swollen scrotum. When I woke up the next day, it was back to normal size! If that weren't enough, I was completely pain free! Since there was no more pain, I stopped taking my pain medication. I was healed and ready for activation!

That same morning, I called my surgeon's office to make my post-surgery activation appointment. There would be no delay. This gave us one more reason to celebrate. It was the best day I had had since coming home from surgery. The infection, swelling, and constant pain were all in the past. Both of us thought the worst was behind us. The song "Happy Days Are Here Again" played through my mind the entire day. After such a great day, we did not anticipate another unanticipated crisis was lurking just around the corner waiting to make its appearance. It was so bad, I actually debated whether or not to include it in this book, but I decided to write about it in the following chapter.

The odds are incredibly slim that what happened next will happen to you if you choose implant surgery. But I decided to include the it for two reasons. First, it was part of my journey with implant surgery. Second, while the odds are low you'll experience the same thing I did, you might face the same temptation at some other point in your life, so it's possible reading the next chapter could save your life.

Questions to consider:

1. Is there something in your medical history or history with surgery that your surgeon should be aware of *before* you have surgery?
2. Discuss the issue of pain control. If you experience significant pain following surgery, have a plan in place for you to obtain pain medication without delay.
3. Are you aware of the signs of infection so you can get antibiotics as soon as possible?
4. Discuss the type of clothing you'll need to wear after surgery, including some form of scrotal support.

CHAPTER 14

Using-Then Abusing—Pain Medication

According to the Center for Disease Control, forty-four people die every day in the United States as a result of a prescription opioid overdose.[1] I'm amazed I wasn't one of those people. I've discovered how easy it is to lie to myself and then to my partner. In the process of writing this chapter, I uncovered the truth about the reasons why something awful happened during my post-surgical recovery.

On the same day my scrotum returned to its normal size, my pain, which was significant, came to an end. Therefore, I stopped taking my pain medication. By the end of the day, I felt as though I were coming down with the flu. I experienced a severe case of the chills, so I put on a winter coat. I knew something was wrong, because everyone else in the house was wearing spring clothing. It didn't take long for chills to stop, and then I began to sweat. At that point I needed to remove my winter coat and change into light clothes. During the next few hours, I alternated between feeling cold and hot.

That night, I was too agitated to stay in bed. I paced around the house and then tried to sit in one place, but I couldn't. As I paced throughout the night, somewhere between four and five in the morning, it occurred to me I wasn't dealing with a sudden case of the flu. Something else was going on. I went online and looked up "symptoms of oxycodone withdrawal."

As I looked at the list of symptoms, I found I was experiencing seven of the twenty most common symptoms:[2]

- Anxiety
- Diarrhea
- Difficulty falling asleep or staying asleep
- Irritability
- Restlessness
- Sweating
- Chills

It was difficult to believe I was going through withdrawal. I kept saying, *This can't be happening.*

Day two was no different from day one. On night two, as I paced around the house, I realized I had reached the end of my capacity to cope with withdrawal from oxycodone by myself. I wasn't sure what to do, but one thing was certain: I would not spend a third night awake, shivering, and pacing around the house. I decided I'd take the minimum amount of oxycodone necessary to end my symptoms. That night, I also decided I was going to call a rehab center the next morning and ask to be admitted for oxycodone withdrawal. I had reached a new low point in my recovery from surgery. I spent the entire night regretting my decision to have implant surgery and dreading my admission to a Detox Center.

I can't put into words my relief when the chills, shivering, and the rest of my symptoms ended on the morning on day three. That night, I went to bed and fell asleep instantly. I had completed the process of oxycodone withdrawal cold turkey. At the time, both Brenda and I were puzzled as to the reasons why I had gone through withdrawal when I had only taken my pain medication for three weeks.

I asked a friend who is a physician how was it possible to become dependent on oxycodone in such a short time. He told me that, following surgery, most people experience gradual pain relief and taper off their medication accordingly. In my situation, I was taking a high dose, and since my pain ended abruptly, I stopped taking the medication abruptly as well. Brenda and I both accepted his explanation as to why I had gone through such a miserable withdrawal process, but the truth remained buried for a few months longer until I began to write this chapter.

I reached the point in my writing where I wanted to assemble a list of the medications I would have taken to a drug detoxification center had it been necessary for me to go there. When I went to the medicine cabinet to pull out everything I had been taking for pain, I experienced a rude awakening. The first bottle I found was my oxycodone. When I decided that taking it every four hours, as instructed on the prescription, was not helping, I started taking it every three hours. Since I was awake a lot, that meant I was taking oxycodone six to seven times a day. Oxycodone can become habit-forming even at regular doses. The higher the dose, the more likely you'll end up addicted.

As I researched the maximum daily recommended dose for oxycodone, I discovered that a total daily dose greater than eighty milligrams, might cause fatal respiratory depression when administered to patients who are not tolerant to the respiratory depressant effects of opioids. I was at a higher risk for that, since I have moderate to severe sleep apnea.

If that's all I did, I would have put my life in danger, but I discovered a whole lot more. Since oxycodone didn't relieve my pain, I went into my supply of prescription medications at home to see what else I could take to reduce my pain to tolerable levels.

The second medication I took was liquid hydrocodone. The maximum daily dose for that drug is six tablespoons a day. I had no way of knowing exactly what dose I was using, because I didn't use a tablespoon to measure it. Instead, I took what I called a "swig." I would lift the bottle to my mouth and drank what I thought was two tablespoons. I did that three times a day.

Each night, I also took tramadol. I never bothered to read the following warning while I took it: "You should not take tramadol if you have used narcotic medications within the past few hours. MISUSE OF NARCOTIC PAIN MEDICATION CAN CAUSE ADDICTION, OVERDOSE, OR

DEATH." Tramadol is another drug that can slow or stop your breathing, especially when used with other pain medication.

Since bedtime was the most painful part of my day, I had the following nighttime ritual for three consecutive weeks. I'd take an oxycodone and wash it down with a swig of liquid hydrocodone. Then I'd take a tramadol. After that, I'd take a Restoril to help me get to sleep. It's reasonable to wonder how on Earth I managed to accumulate so many different pain medications. The truth is, I hoarded my pain medications. I stopped taking them as quickly as possible so I could save them for another bout with chronic pain. That was the first time in my life that I had ever combined pain medications in this way.

As I looked up each of the medications I was taking, I stopped wondering and gave thanks that I had survived the potentially deadly combination of powerful pain-relieving prescription drugs that I had been using daily for three weeks. To this day, I'm amazed I wasn't one of those accidental overdose statistics. If I had brought all of the prescription pain relievers I was using to a detox facility, I have no doubt they would have classified me as a serious abuser of pain medications who had developed a potentially life-threatening habit.

What I learned about myself isn't pretty. I suspect this applies to tens of thousands of people who experience unrelenting and/or chronic pain. To experience pain relief, I was willing to lie to myself and to my wife. I was willing to risk the possibility of addiction, drug dependence, an overdose, or death. I was single-minded. I needed pain relief, and I was willing to do anything to bring my pain down to a manageable level. I never expected that one of the most important takeaways from my experience with penile implant surgery would be an up-close and personal understanding of how people end up addicted, suffer an overdose, or die as a result of mixing a variety of pain-relieving drugs.

Once I gained those insights, I knew I needed to tell Brenda the truth, that I had been combining a variety of prescription drugs without telling her. Since the danger had passed and I had survived, I seriously underestimated the devastation Brenda experienced when I told her what I had done. I assumed that she would be as grateful as I was that I had survived. Brenda wasn't grateful at all. She was rightfully ticked off. I had lied to her. I'd taken dangerous and potentially fatal amounts of prescription drugs, and I had become drug dependent.

I went through three days of withdrawal without Brenda knowing that I had combined and abused multiple pain medications. By far the worst consequence of confessing to Brenda how I had abused multiple pain medications was the loss of her trust in my judgment. Brenda's greatest fear is that I'll repeat this behavior of lying and combining pain medications the next time I face chronic pain. It's important to me and her I don't make that mistake a second time. Given my age, medical history, and all the medications I had combined, there's a high probability I might not survive the next time I do something that stupid and dangerous. Since that episode, I have agreed to tell Brenda when, how much, and how often I'm taking pain medication in every circumstance.

After reading about my hospital experience and recovery at home, you might be discouraged and say to yourself, "There's no way I'm going through surgery." Remember that most of my experiences were unique.

If you don't have interstitial cystitis, it's doubtful you'll experience bladder spasms. It's also improbable you'll stop producing urine. If you have a drain put in during surgery, it's unlikely your scrotum will swell to the size of a grapefruit. If you require prescription narcotics, you know you'll need to taper yourself off rather than quitting cold turkey. In addition, you've learned that it's a bad and potentially fatal idea to combine pain medications.

A few days prior to editing this chapter, I was reminded of the danger of using and abusing pain medication. The death of the rock star Prince was all over the news. Apparently, Prince died from an overdose of prescription pain medication. It's another reminder that abusing pain medication is a serious and potentially deadly mistake. Based on the combination of drugs I took, I have no idea how I survived.

If you go down the surgery route, your journey might go perfectly without any incidents out of the ordinary. You could also face your own unique challenges. There are no guarantees when you decide to have a surgery of any kind. I'm certain that by reading about my experiences, you have reduced the odds that you will suffer through the majority of the issues I experienced.

Thankfully, going through withdrawal was the last crisis I experienced on my journey—well, not quite. Now that my scrotum was back to normal size, I still had time to do something foolish before my activation date. I began toying with the idea of cheating and using my implant before my activation appointment. I think that is a common temptation.

It was easy to find the bulb I needed to pump. I was also convinced I had found the release button, but I was wrong. In reality, I had no idea where the release button was located. During my many attempts to squeeze the pump, I found it too difficult. Even though I squeezed the bulb inside my scrotum as hard as I could, nothing happened. I thought of wrapping my scrotum in a towel and using a vice grip, or pliers to pump myself up. I ruled out each of those ideas since they had the potential to cause serious pain. The fact that I took the time to visualize how I could use each of those tools to inflate the pump speaks to my high motivation to use the implant before my scheduled activation date.

Once I get an idea into my head, I'm persistent (my detractors would say "obsessed"), so multiple failures didn't cause me to give up. I went online and discovered that hot water makes it easier to inflate the pump. As luck would have it, we have a hot tub. So I went into our hot tub with Brenda to squeeze the pump. At the time, I considered each squeeze a victory. Here's what I didn't know: After surgery, most men are sent home partially inflated/erect. I didn't realize that, because I attributed my larger size to the fact I had inflatable cylinders placed on each side of my penis.

With a few successful pumps, I found myself fully erect. Sweet success! We left the hot tub to enjoy our first sexual experience with the implant. Since both of us experienced a significant amount of guilt, worry, and concern that using the implant early might cause some damage or delay in my healing, so we decided to enjoy just one encounter. Okay, we had two. Both experiences were fantastic!

When it was time to deflate the implant, I pushed what I believed to be the release button. Nothing happened. That should come as no surprise, since I wasn't really pushing the actual deflate button. After multiple attempts to deflate the implant failed, I gave up. It took a few hours before I realized that living with a non-stop erection is both uncomfortable and painful.

For the next few days, I attempted to deflate my implant. Nothing I did was successful. There was no way I was going to call my surgeon in San Francisco to ask for an emergency appointment. I'd be forced to admit to two things I didn't want him to know. First, I had used the implant earlier than I was told. Second, I had no idea how to deflate it. I chose to suffer in silence instead. It never ceases to amaze me how much I'm willing to suffer to hide my embarrassment or shame. Looking back, it was a foolish choice. Had I called for an emergency appointment, I'm sure my surgeon and I would have laughed together about my predicament. Thinking rationally, I'm sure he's dealt with such issues dozens of times.

Most of the men who choose implant surgery have experienced erectile dysfunction for years. It's understandable they are eager to take our new equipment on a test run as quickly as possible. I discovered that a few moments of sexual pleasure were definitely not worth the three weeks I spent erect, uncomfortable, and in pain. My best advice is to wait until your activation appointment before you test out your new equipment. And if you need help, ask!

Questions to consider:

1. Do you have a plan to deal with chronic pain in case that's your experience after surgery?
2. Do you have any history of alcohol or drug abuse that might impact your judgment when seeking relief from pain? If so, who

will help you avoid self-destructive behavior and/or life-threatening behavior?

3. Do you have a tendency to follow or disregard the advice of your doctors?
4. How will this tendency manifest itself with your post-operative instructions to wait six weeks before you use your implant?
5. How do you expect your implant will affect your self-esteem?
6. How will the return of your erectile function affect your sexual relationship and behavior?

CHAPTER 15

Activation Day

Some men find their implant activation painless. Others experience mild to moderate discomfort. I decided to minimize any pain by taking an over-the-counter pain-relieving medication two hours before my activation appointment. Once I was on the examination table, my surgeon told me to pull down my pants. I experienced a significant amount of embarrassment when he demonstrated how to use the implant by squeezing my scrotum to inflate it. He also demonstrated how to push the release button in my scrotum while simultaneously squeezing the base of my penis to deflate my implant. He had to show me it twice before I was ready to try it on my own.

I was surprised how much easier it was to squeeze the pump once it had been cycled the first time. Since that office visit, I've experienced one painful mistake multiple times. While pumping, I've lost my grip on the pump, sending it crashing into a testicle. That results in brief but severe pain. To avoid that painful experience, I've learned to use both hands when I pump. I use one hand to squeeze the bulb and the other to hold the base steady. I've never lost my grip on the bulb with the two-handed approach.

When it came time to deflate the implant, I was unable to locate the button. My urologist had no problem, and neither did Brenda, but try as I might, I could not find it. My inability to do so became a source of embarrassment. After I was inflated for the third time, I was the only one in the room who couldn't deflate the implant. To this day, I'm grateful Brenda came with me to my activation appointment. Once I was certain she knew exactly how to find the button and use it, I was done. I didn't need to fail one more time. I was ready to get out of the examination room and have Brenda teach me in the privacy of our bedroom rather than lying on a table with my pants down and my surgeon by my side.

It turns out there was good a reason why I couldn't find the deflate button. It wasn't where it was supposed to be. My pump was not lying in the expected position in my scrotum. It rests on a forty-five-degree angle. That means the deflation button is on the side rather than the front of the bulb. I could not picture that in mind. I needed Brenda to hold the practice bulb by my scrotum so she could demonstrate the exact angle of the bulb and the location of the release button. From that day on, I could deflate the implant by myself.

As I reflected back to my activation day, I made the following observations and recommendations:

Ask your surgeon if taking an over-the-counter pain reliever to minimize the pain of activation is a good idea *before* you take it.

- The pump starts off difficult to squeeze and takes a lot of pressure. I was only able to get four to five pumps before it got too hard to pump any more. Pumping gets much easier over time.
- Even though I was warned, I was surprised to discover how small my erections were—shocked was more like it. .
- If you go for activation by yourself, make sure you don't leave the office until you can inflate and deflate your implant. I let my embarrassment take over, so I left the appointment without learning how to deflate my implant. Don't make my mistake if you are alone for your activation appointment. Take as much time as you need until you are positive you can inflate and deflate your implant.
- It's impossible to over-inflate your implant, however, it is possible to inflate it to a point where you experience pain. In the first few

> months, in an effort to increase the size of my erections, I pumped until I experienced pain.

Pumping until I experienced pain taught me three valuable lessons. One, pain seriously diminished my enjoyment and pleasure. Two, Pain frequently took away my ability to achieve an orgasm. Three, pumping until I caused myself pain didn't add any degree of pleasure for me or for Brenda. All of this meant it was far more important for me to remain comfortable and pain-free than it was for me to attempt to get a larger erection.

Using an implant is an easy process. What isn't as easy is deciding how to use it before, during, and after sex. I've known some single men who don't tell the person they are dating they have an implant. They make a trip to the bathroom before sex to pump themselves up prior to intercourse. I have no idea how they explain their on-going erection once they've experienced an orgasm. Obviously, married couples do not have that dilemma. I prefer to have Brenda rub me as I'm pumping so that both of us can feel me becoming hard at the time she's providing stimulation.

At first, it was important for me to deflate my implant following an orgasm so it would feel the same as it did before surgery, excitement followed by relaxation. Now it doesn't matter if or when I deflate my implant. Sadly, one of the hazards of getting old is there is an ever-increasingly number of things you forget on a daily basis. More than once, a few hours after sex, I've experience pain in my groin. When I've searched for the reason why, I discovered that I forgot to deflate my implant! That has happened enough times now that I'm beginning to find humor in my forgetfulness.

I think both of us were shocked with the sudden and immediate changes brought about by activation day. After a few days of use, Brenda and I were enjoying our sexual experiences together as much as we did before my prostate was removed.

I found the implant to be light years better than the vacuum pump, penile injections, or ED medications. My success rate with the implant is one hundred percent. Failure to achieve an erection is not a possibility, worry, or concern. I can't even put into words the change that occurred in my level of confidence. It's one of the amazing psychological and physical gifts of having an implant.

I feel completely restored. Brenda and I look forward to many years of making love. The decision to have implant surgery changed my life and my enjoyment of sex for the better. It was the best decision I've made since my prostate was removed.

Questions to consider:

1. Does your surgeon suggest you take any over-the-counter pain medication before your activation?
2. Do you want your partner to attend your activation appointment?
3. What can you plan together to make your first experience with your implant a romantic time?
4. After a few weeks, discuss the positive and the negative changes the restoration of erectile function has brought to your relationship.

CHAPTER 16

Sometimes Survival Isn't the Best Choice

Before becoming the "new me," I had to deal with the old me. Throughout life, each of us learns and maintains a series of old survival strategies. These strategies can lay dormant for decades until a traumatic event or a crisis serves as a trigger to activate our old and familiar survival strategies.

Our survival strategies are highly valuable to us, because they helped us endure a difficult event or era in our lives. When we utilize our old survival strategies in a present-day crisis, without realizing it, we take a journey back in time. As a result, we experience our new crisis with the skills, filters, coping mechanisms, and ways of thinking we developed when we first created and utilized our old survival strategy.

Dealing with the trauma of ED with your old survival strategies increases the likelihood you will cope in ways that will not produce healthy growth and change. It took four years before I understood which of my old survival strategies were triggered as I coped with my ED. Two strategies in particular seriously wounded me and my wife and damaged the fabric of our marriage.

The first survival strategy I used I refer to as my "I'm all alone in the world" strategy. This strategy has caused nothing but trouble in my marriage. I developed this strategy when I was eight years old. A neighbor of mine, known to be a bully, threatened to harm me physically. I took his threat seriously and ran home as frightened as an eight-year-old can be. My mother was in the basement ironing. I told her what happened. I expected to receive some type of assistance or the reassurance of parental protection. Instead, she totally ignored my fears.

I remember walking up the stairs thinking I was only eight years old, but I was on my own. At the tender age of eight, I decided my mother was useless in a crisis. I concluded that to survive, I had to learn to fend for myself. With each and every crisis I faced thereafter, I'd go back to my survival strategy, which began with the belief "I'm all alone in the world, and I need to fend for myself."

Brenda can attest to the number of times we faced a crisis together but alone. That's the unintended consequence of utilizing survival strategies from our past. They helped us survive before, but they cause harm in the present to ourselves and/or to those closest to us. Every time I fell back on my survival strategy of coping with a crisis alone, I shut Brenda out. When I shut Brenda out, we lost each other's comfort, support, strength, encouragement, and wisdom. Unfortunately for both of us, ED activated my "I'm all alone in the world" coping strategy.

A second strategy has plagued me for decades. It developed when I was twenty-one. That was when my father decided to leave my mom after twenty years of marriage. How he left created a host of destructive ways of thinking about marriage.

It was a typical day in the life of my parents. My mother and father woke up, shared breakfast, and went work. On that fateful day, my father came home from work early in the afternoon. He packed his luggage, placed his bags into his car, and drove away. He left my mom without notice or warning.

When my mom returned from work that evening, she discovered her husband of twenty years had deserted her. All she found was a note from my father, which was filled with lies. He told her that he was taking a brief trip to find himself. He said he expected to return home within a short amount of time. His note didn't contain a single shred of truth. He wasn't going on a trip to find himself, and he had no intention of coming back. He was heading to Florida to start a new life with a woman he had met while taking a solo vacation to Florida.

What made it so devious was that a few months prior to leaving, my father agreed to go into counseling with my mother for the purpose of working on their marriage. This was one more lie. He had already decided to leave. His hidden agenda was to have my mom in therapy so she would have a professional to talk to after he abandoned her.

As a result of this event, I developed an unshakeable belief that love and marriage could never be trusted. My first marriage ended in divorce within a year after my wife cheated on me. In addition, for some odd reason, most of my male relatives shared the details of their extramarital affairs with me. I grew up thinking every married man cheated on his wife. I concluded it must be true for their wives as well. After my father deserted my mother and my own marriage ended in betrayal, I developed the unshakable belief that marriage was a temporary relationship that ended with an act of infidelity.

When I married the second time, I believed I had found a woman who would be faithful. But the longer Brenda and I were married, the more anxious I became. From my perspective, it was inevitable that betrayal and divorce were just around the corner. I exhaled for the first time when we reached our twenty-first wedding anniversary.

Once I became impotent, I was absolutely certain Brenda was going to betray and leave me. The only thing I didn't know was which event would occur first: the betrayal or the departure. Brenda could do nothing to alter my expectations. I was in my "prepare to be deserted" survival strategy. What added to my pain was the fact that I believed I deserved to be deserted.

Brenda was stuck with a defective man. As much as I wanted to stay married, I believed she deserved to be married to a fully-functioning man. I bided my time until Brenda grew weary of living with me, a man with erectile dysfunction. I expected my marriage to end in the same way my parents' marriage had ended. One day I'd come home from work and find a letter on the table informing me she had left me for another man. If Brenda hated erectile dysfunction half as much as I did, how could I blame her for leaving me for a man who was capable of performing in the bedroom?

The reality was, Brenda never considered leaving me, but I couldn't believe that, because I wasn't living in the present. I was stuck in the past with my "preparing to be deserted" filter fully activated. That survival strategy filtered all of my thoughts, beliefs, feelings, attitudes, and expectations. Since I was stuck in the past, there was no way I could view my present situation accurately.

My two survival strategies and their filters created a perfect storm in our marriage. Without talking to Brenda, I decided it was vital to my survival that I adjust to living as if I were alone. I was convinced it was only a matter of time before Brenda would write me a goodbye letter. From my perspective, it made complete sense to shut Brenda out of my life. We lived the first eighteen months of our marriage as cancer survivors together, alone. That's the destructive nature of old survival strategies.

The reality was I had a loyal, faithful, and loving partner who needed my attention, affection, love, and support so we could face this crisis together as a team. My old filters prevented me from adjusting my beliefs and behaviors so I was free to respond to the faithfulness and love Brenda was offering. As I prepared to live alone, I abandoned Brenda emotionally, relationally, and physically. I abandoned myself as well.

That provides you with a picture of how my past derailed my capacity to cope with ED in any way that would bring us closer together as a couple.

Most men with ED face several present-day issues as well. Three close companions of ED are:

- The loss of your sense of manhood
- The loss of confidence in the bedroom
- Toxic shame

The most common cry of pain I hear from men who experience ED comes from the belief that ED has taken away their manhood. I was surprised to discover I was among those men who believed that lie. If you had asked me prior to ED entering my life if my ability to have an erection made me a man, I would have thought the question absurd. I would have denied the connection between my ability to have an erection and my sense of manhood. I discovered there was, in fact, a deep connection between the two. When I became impotent, I believed I was no longer a man.

Any time Brenda and I were in the bedroom, I was fairly certain I would not achieve an erection hard enough for penetration. That constant disappointment and failure led to a total loss of confidence. When someone experiences failure after failure it's only human to want to avoid the situation that brings this sense of failure and disappointment.

The complete loss of confidence in the bedroom was a result of my laser focus on my erectile dysfunction. The truth of the matter was that my wife and I had discovered other ways to please each other. While Brenda shared my disappointment when I could not obtain an erection, she didn't share my sense of failure or shame. She found pleasure and enjoyment in our sexual relationship. All I felt was an overwhelming sense of toxic shame.

Toxic shame is the least understood companion of men and couples coping with ED. It is also the most destructive. Toxic shame devastates a man's self-esteem. It speaks condemning words that drive men into hiding from their partner and themselves. Toxic shame drives away the possibility of hope and replaces hope with the painful state of hopelessness and despair. Believing they are worthless, men become depressed. When women are depressed, they typically experience sadness. Men who are depressed typically become irritable.

Angry words and/or isolating behaviors replace all forms of affection and love. Under the influence of toxic shame, there are no redemptive opportunities. To make matters worse, two things happen. First, men are driven to find relief from toxic shame with mood-alternating behaviors. They seek out alcohol, drugs, TV, computers, pornography, or other escapes to reduce the pain. Even though those escapes provide little or no relief, men use them over and over again.

There's a healthy choice to make. Unfortunately, most men under the influence of toxic shame refuse this option. To defeat toxic shame, you must come out of hiding and seek outside help. Just the thought of seeking help brings up feelings of anger and defensiveness. There's no way you are going to tell anyone about your suffering with ED. If you find yourself alone and isolated, you've given toxic shame a complete victory.

I invite you to come out of hiding. To achieve victory, you need to reconnect with yourself and your partner. You need to confront and challenge your toxic shame. To do it successfully, you need assistance from men further along in the journey with ED. As I've mentioned, a great place to find them is in an online forum, where you don't have to reveal your true identity.

Coming out of hiding, reconnecting with yourself and your partner, and joining a community of men are all good first steps. Three more difficult steps might also be necessary. The first is consulting with a physician to determine the cause of your ED. Without knowing the cause, finding an effective treatment is impossible.

The second step involves damage control. That might require professional help. You may have caused serious damage to your relationship with yourself and/or your partner that will require outside professional help to repair. I know I needed that help, and you might need it, too. As Brenda and I assessed the damage of my coping strategies, I discovered she felt rejected, unattractive, unappreciated, and unloved. I had caused so much damage to the fabric of our marriage it was necessary for both of us to seek professional help. I regret how long it took for me to agree to get help

The third and final step toward healing involves grieving the losses of your old sexuality and discovering a new sexuality. I got stuck. Anger and bitterness prevented me from grieving. I hated my post-surgery sexuality. Here is a list of losses, some permanent and others I hoped were temporary.

- Initially, surgery took away my capacity to ejaculate. For a long time, I doubted if I would ever enjoy an orgasm again.
- The intensity of my orgasms diminished to the point where, at times, I didn't know whether I had experienced one.
- I experienced climacturia (urination during orgasm) frequently.
- My interest in sex was high throughout my marriage. After I
- was diagnosed with cancer, my interest in sex vanished completely.
- The fact that I couldn't get an erection caused me to feel awful as a man. All sexual activity became nothing more than a bitter reminder of my manhood.
- I believed every condemning voice emanating from my toxic shame. Since I don't enjoy feeling like a failure, I avoided sex and all forms of physical affection with my wife.
- The physical pleasure associated with sex and affection was completely lost.
- Each and every sexual encounter left me feeling disappointed, angry, ashamed, and defective.
- The loss of our affectionate touching led me to feel lonely and unloved.

It's highly likely you have your own personal losses and survival strategies. I encourage you to write them down and ask yourself if you are willing to face the pain of these losses one by one and the unintended consequences of your survival strategies. For far too long, I avoiding taking that step, so I remained stuck in a process of avoiding anything that would trigger my feelings of loss. I did not want to face those painful losses or go through the healthy and necessary grief process. Since sexual activity triggered my losses and painful feelings, I avoided all expressions of my sexuality and all expressions of Brenda's sexuality.

My quest to avoid all forms of sexual expression caused a number of serious problems in our marriage. My use of old survival strategies also caused an increase in fighting and marital tension. We were in a standoff of differing beliefs. I believed Brenda would be better off with a real man who had the capacity to have an erection. Brenda believed it was possible for us to maintain a mutually exciting sex life. She also knew I hadn't lost my ability to be affectionate, loving, and caring.

Even though it was an emotionally lonely and painful time for Brenda, she was ready, willing, and able to confront me and fight for our marriage. I'm grateful she didn't give up on us, because I did. As a result of her persistence and her belief in me, I was motivated to challenge my old survival strategies and begin the grieving process to make peace with my post-surgery sexuality. I had some success, but I never accepted my losses as permanent. I continued to hope for nerve bundle healing.

Four years after surgery when my healing was over, it was obvious I had not made peace with ED. I was biding my time, waiting to heal. Once I learned my ED was permanent, I realized I did not want to live the rest of my life without the ability to enjoy physical intimacy with my wife It became clear to me I had to do something to restore my ability to have an erection. At the time, had no idea that implant surgery was the option that would bring back my erectile function. Surgery is not the right option for every man coping with ED Living and loving in healthy ways is an option for you no matter what you decide.

If you decide to live with, ED I don't have to tell you there are unhealthy ways to think about yourself and your sexuality. If you've lived with ED, you've experienced a season of suffering emotionally, relationally, and sexually. Here's a sample of common but unhealthy and false beliefs that magnify that suffering:

- Without the ability to have an erection, I'm a failure as a sexual partner.
- My partner will find sex with me to be a miserable and unsatisfying experience.

- I don't deserve to be in a relationship if I can't satisfy my partner with an erection.
- I want to avoid feeling like a failure, so I'll avoid all sexual experiences.
- It's impossible for me to enjoy my sexuality without an erection.
- It's impossible for my partner to enjoy my sexuality without an erection.
- It's impossible for my partner to enjoy their sexuality without me having an erection.

You can add your own negative, pessimistic, judgmental, and harsh beliefs about yourself and your partner to this list. Toxic shame would have you believe there's no way for you to enjoy your sexuality and your partner's sexuality. In addition, toxic shame makes you believe there's no way your partner could enjoy your sexuality or their sexuality with you. Based on these false beliefs, you come to the conclusion you are doomed you to live a lonely, sexless life. Do you want to spend the rest of your life under the power of toxic shame?

It might be difficult for you to believe, but many men live a healthy and exciting sex life after losing their ability to have an erection. Some couples would say their sex life got better afterwards. It is beyond the scope of this book to help men and couples find the various ways they can learn to please and pleasure each other, but I want to provide some resources for those who have decided to live with ED.

- Great Sex Without Intercourse: HYPERLINK "http://www.aarp.org/home-family/sex-intimacy/info-12-2012/great-sex-without-intercourse.html"
- Keeping Sex Alive When You Face ED: HYPERLINK "http://tolovehonorandvacuum.com/2014/04/ed-in-marriage/"
- A Six-Step Survival Guide For Women: HYPERLINK "http://www.phoenix5.org/companions/6steps.html"
- Readjusting after Erectile Dysfunction: HYPERLINK"http://chealth.canoe.com/channel_section_details.asp?text_id=1588&channel_id=16&relation_id=27919"

Implant surgery is not the right choice for every man facing a lifetime of impotence. There are healthy and unhealthy reasons to decide against implant surgery.

Whether or not you have implant surgery, I suggest you take the time to face the emotional, relational, and sexual impact that ED has had on both you and your partner. The gift that comes from ED is that it can move you into greater emotional health. You can challenge your old survival strategies. You can develop new and healthy attitudes and loving behaviors

Remember that you haven't lost your manhood, you've lost your erectile abilities. There's a difference. As you discover new ways to express romance and enjoy your physical relationship, you'll develop a new definition of manhood. Your new perspective will enable you to challenge and defeat the lies that come from toxic shame. This isn't easy to accomplish, so get whatever help you need to say good riddance to toxic shame. For those men who have decided to live with ED, it doesn't mean it's the end of your ability to give and receive pleasure. It's up to you and your partner to discover new ways to enjoy sex together.

Even though I chose an option that restored my erectile function, it was obvious to both of us that an implant wasn't going to erase the pain of the last few years. I had caused so much damage to the fabric of our marriage it was necessary for both of us to seek professional help. If you've been fighting, using drugs or alcohol to cope, and/or you are avoiding physical and emotional intimacy, I urge you to seek out professional help. I regret how long it took for me to agree to get help, but I'm glad I followed my wife's prompting to seek out professional help. We were unable to repair our relationship by ourselves.

We've come a long way! I am writing this chapter in Cocoa Beach while Brenda and I enjoy a romantic vacation. Brenda is sleeping peacefully as I write. We've experienced an amazing and restorative time together. This vacation reminds me of our honeymoon. In many ways, it's better. I've learned so many important life lessons as a cancer survivor. One of lessons was that ED doesn't take my manhood away unless I allow it to do so. It's sad for me to admit, but most of our romantic vacations we took after my diagnosis of prostate cancer ended in disappointment, fighting, and failure. The mistake on my part was defining success as achieving an erection that was useable for penetration. Anything else caused me to react with disappointment, anger, frustration, and shame. I felt as though I was worthless as a lover and as a husband. I regret that my attitude and ignorance prevented me from making peace with ED. You don't need to like being impotent to achieve peace and make the best of what you can do.

You don't have to be a cancer survivor to realize that our time on Earth is limited and too valuable to waste. Holding on to grudges, fighting constantly, and living with an unforgiving heart or broken relationships is a colossal waste of time and energy. That might sound corny, but it's true. Love is what life is all about. Loving God, loving others, and knowing how to love yourself are the things that give life meaning and purpose.

When you believe you're unlovable, you'll start depending on things outside of yourself to make you lovable. Some depend on youth or beauty, others on money, power, or success. Most men depend upon their erectile abilities. You don't realize you've made that decision until you've lost your ability to maintain an erection. That's when you realize you've lost a lot more than your ability to have an erection. You feel as though you've lost your manhood and value as a man. The reality is an erect penis does not make you a man.

Erectile dysfunction can teach you valuable lessons about genuine manhood if, and only if, you have a teachable spirit. I didn't choose to have a penile implant to reclaim my lost manhood, though some men do. I chose to have a penile implant because I missed the physical and emotional closeness of my wife and I becoming one in the bedroom.

We live in an amazing medical era. What was lost to both of us has been restored. With that restoration has come a sense of all-encompassing gratitude that gives added meaning, value, and importance to the time we spend together, whether in the bedroom or holding hands as we walk along the beach. We've had an amazing journey coping with prostate cancer and erectile dysfunction.

Last night, we received a phone call inviting us to be keynote speakers at a cancer survivors' conference. Public speaking is not an activity that's in my comfort zone, neither is writing this book. My faith and cancer are powerful reminders to live life fully. This requires me to spend a lot of time outside of my comfort zone.

I'm grateful for the life lessons I've learned as a result my experiences with prostate cancer and erectile dysfunction. Whether you decide to live with ED or go for implant surgery, you remain a man, fully capable of giving and receiving love. I wasted years of my life believing otherwise. If you make the same mistake I made, believing that it's impossible to be a man or to enjoy a mutually satisfying sex life without your erectile abilities, I say to you, "Danger, Will Robinson!" You are about to sentence yourself and your partner to a lifetime of misery. It is vital for your sexual, emotional, and relational health to challenge this commonly held belief.

Questions to consider:

1. Can you identify any of the survival strategies you are using to help you cope with ED?
2. How have those strategies caused conflict, misunderstanding, or misery in your relationship with your partner?
3. Ask your partner to answer questions 1 and 2.
4. Do either of you think you would benefit from professional help? If either one of you answers "yes," my advice is for *both* of you to get help together.

CHAPTER 17

Post-Implant Surgery: A Wife's Perspective

It was the dawn of new day! All of Rick's pre- and post-surgery issues were resolved. We were both excited to experience each other physically again in a more complete way. We approached the new sexual experience cautiously. Both of us had vivid reminders of the many disappointments we had experienced along the way.

One day while sitting in the hot tub, Rick told me he managed to pump his implant a few times and that he was hard enough for penetration. Rick wanted to do a trial run prior to his activation appointment. I was reluctant to do so without proper knowledge and clearance from the doctor. Still, the moment presented itself, and it was amazing. Having something I had lost restored was so delightful. I had spent so much time grieving the end of our sexual relationship. I thought the pleasure of making love was lost forever. The implant surgery gave us the ability to enjoy that part of our sexuality once again. It brought me great happiness. It also restored our tenderness, and I felt such relief and overwhelming joy that I cried.

For the next few weeks as we waited for Rick's activation appointment, I wondered if the implant was really going to bring my husband back to me both emotionally and physically. Continuing our work on our relationship was a positive step for us. We were learning to approach each other in a kinder and more loving way. We were both trying to learn new things about each other. It changed the environment between us in a positive way. I learned to voice complaints without putting down or criticizing Rick. We shared more of what we appreciated about each other each day. Becoming more playful was a great foundation, which set the stage for a new phase in our relationship.

On the eve of Rick's activation day, we reserved a hotel at Half Moon Bay. We had a room overlooking the ocean. It was the first of many romantic vacations together.

As we headed to San Francisco six weeks after surgery for the activation appointment, a number of thoughts crossed my mind. I wondered if I was I capable of performing and enjoying sex with an implant. I wondered if my experience would feel mechanical rather than natural. I wondered if I would feel differently, since I wouldn't be the one to give my husband an erection. Erections were now the result of the implant. My sexuality was no longer needed. It felt like a huge loss to me that I needed to grieve. I wondered how different sex would feel with an implant and whether it would interfere with my own pleasure. I also wondered if I'd feel uncomfortable pumping the implant in front of Rick's surgeon.

Finally, activation day arrived. The surgeon asked Rick to pull down his pants. He squeezed the pump in Rick's scrotum until Rick was fully erect. Then he deflated the implant. Next, it was my turn. The training was embarrassing, but I practiced a few times. It did take a bit of "going blind," as the inflate and deflate procedure requires the use of touch without a visual cue. Since the pump wasn't lying correctly in Rick's scrotum, locating the deflate button was a bit of a challenge. As we practiced in the office multiple times, I remember thinking that I hoped I wasn't hurting Rick, as he had been through so much pain and swelling in that area.

It didn't seem to hurt him at all, and I became proficient quickly with both inflating and deflating the pump during the office visit. As soon as Rick was confident in my abilities to pump and deflate the implant, he wanted to end this appointment as quickly as possible. I don't think he felt comfortable with the surgeon or me pumping him erect in the surgeon's office. As we left the office, I wondered if I'd be able to do it without the supportive surgeon at my side.

There was quite a contrast as we left our activation appointment compared to when we had left that same office after being told that Rick would be impotent for the rest of his life. This time, we were leaving with joyful anticipation, but both of us also experienced anxiety and fear. In the years after prostate surgery, we had taken many romantic vacations where Rick would not respond to his ED medication. Those vacations resulted in disappointment, frustration, and fighting. It was difficult for us to relax and believe the days of failure and disappointment were coming to an end. While our PTSD symptoms were improving, it was easy to fall prey to destructive thinking that led me to feel stress and anxiety, but we were determined to keep going.

It was a relatively short drive from San Francisco to the inn at Half Moon Bay. Finally, Rick and I were on our own. I thought I needed to take things slowly. I did not want pain to be part of the experience. I was reluctant to act aggressive sexually, as that had led to frequent fights and rejection during the years Rick was impotent. For the sake of my emotional safety, I needed Rick to take the lead. He was willing and now physically able to do that. We made it work! Yay for us! Our experience together was wonderful!

I did miss the opportunity of arousing him to erection. We solved that problem by including me in the inflation process. That was sufficient for the moment, but later, I found it to be a significant loss that I needed to grieve. For that wonderful, romantic trip, I put that issue aside. We were making love again, and it felt so sweet and natural. It was the most natural method of promoting our intimacy that we had tried over the previous four years.

It wasn't difficult to say goodbye to the vacuum pump, penile injections, ED medications, and testosterone. At Half Moon Bay, we had a good laugh when I discovered Rick had not learned how to deflate his implant. He was so embarrassed he couldn't figure out how to deflate his implant that he hadn't let on during the office visit. I understood his embarrassment. We both imagined what would have happened if I hadn't learned how to deflate his implant. Imagining a trip back to San Francisco to get deflated had us both in stitches. Laughter and humor are healing to the soul.

We've been enjoying our sexual relationship with the implant for almost a year now. Our sexual relationship has blossomed to a point beyond what we could have imagined beforehand. We have grown creatively and learned how to love and cherish each other more than ever. We continue to improve the way we relate to each other, which enhances our sex life. The implant has renewed and revived our sexual relationship way beyond our wildest imaginations. We had a very good sex life prior to the prostatectomy, but the penile implant brought us to us to a whole new level. Even through the darkest moments, storms, and bombs, I am glad to love my man with implant surgery. It has provided a new way of life for us both.

Questions to consider:

1. What relational issues are important to discuss before your sex life is restored?
2. What new romantic ideas or plans can you come up with to celebrate your newfound abilities?
3. How do you anticipate the restoration of your sex life with affect your relationship?

CHAPTER 18

Single with a Penile Implant

Little information is available for single men who choose implant surgery. The information I found I gleaned for this chapter comes from posts made by single men on penile implant forums.

The most pressing question that came up repeatedly was this: When do you tell the person you're dating that you have a penile implant? The answers provided to this question varied. Some men felt the need to disclose the information on the first date. Other men decided to keep it to themselves even after the relationship evolved into a sexual relationship.

The motivation to keep the information private varied. Some men were so ashamed that they kept it a secret to avoid embarrassment. Other men wanted their partner to be impressed and amazed with their ability to maintain an erection, so they kept their implant surgery a secret.

There's no one right answer to this question. Personally, I'm not a fan of deception. I don't think that's a great way to begin a relationship. As far as the timing of when to tell someone about your implant, that's a decision you need to make for yourself.

I believe both single and married men with implants face a dangerous temptation. The restoration of erectile abilities results in a rapid and amazing psychological transformation. One day a man is depressed, feels as though he's no longer a man, and like he has nothing to offer in a relationship. After activation day, his manhood and self-esteem is restored. For some men, that transition can lead to disastrous or self-destructive behavior.

Candy Spelling, wife of famed TV producer Aaron Spelling, had the following to say about dating a man with a penile implant: "My bionic man could go on for five or six hours, and there is no woman, middle-aged or otherwise, who wants to have sex for that long. It was like running a marathon." She broke up with Mr. "Pump and Dump," as her girlfriends had cheekily nicknamed him, because he was "getting too attached," and she just couldn't stand "those six-hour romps anymore."[1]

If this report is accurate, it's an example of a man who became incredibly selfish and self-centered. Perhaps he became an exaggerated version of who he was before his implant surgery.

This leads us to an important question. What's your purpose for implant surgery? Here are a few possibilities:

- To chalk up conquests on your belt buckle
- To increase your value in the dating marketplace
- To feel closer to the person you are dating
- To restore what was lost as a result of erectile dysfunction
- To satisfy a biological and physical need

Kelli Miller has been covering health and medicine for nearly 20 years. She is a board-certified editor of the life sciences. She found twenty reasons why people have sex.[2] These include:

- Duty
- Enhancement of power
- Enhancement of self-concept
- Experiencing the power of one's partner
- Feeling loved by your partner
- Fostering jealousy

- Improving reputation or social status
- Making money
- Making babies
- Need for affection
- Nurturance
- Partner novelty
- Peer pressure or pressure from partner
- Pleasure
- Reducing sex drive
- Revenge
- Sexual curiosity
- Showing love to your partner
- Spiritual transcendence

The return of erectile function restores the opportunity to enjoy sex that involves penetration. The decisions you make with regard to the meaning of sex and your motivation for engaging in sex will determine whether or not you will enjoy and experience blessings or suffer emotional, relational, and sexual harm as a result of regaining your erectile function.

Growing up in the 1970s, sexuality promiscuity was encouraged. My dad advised me to, "sow your wild oats while you're single." He believed the more sexual partners I had before I got married, the easier it would be to settle down once I got married. His advice made perfect sense to me, so I had dozens of sexual relationships prior to getting married.

It took many years for me to discover his advice was seriously flawed. As I observed the effects of living out this advice, my father didn't do very well. In two of his three marriages, he cheated on his wife. We never discussed whether he found a way to remain faithful in his third marriage, but I doubt he did. "Sowing his wild oats" did nothing to help him to remain faithful in his marriages.

I also found out later that there was a high price to pay for my season of sexual promiscuity. The damage of engaging in sex with multiple partners remained hidden until I was married. Sex lost its meaning. For decades into my marriage, I still viewed some of the women I met as potential sexual partners. I wondered if other men struggled in a similar way. I was too embarrassed to talk about it, and I don't know if I had the ability to put the trouble I was experiencing into words.

Knowing my father's history and my sexual past, Brenda said something to me that was the only thing she could have said to keep me faithful for the next thirty-five years. Before we were married, she warned me that if I ever cheated on her, there would be no forgiveness. She assured me that one episode of stepping out of our marriage sexually would result in the destruction of our marriage by divorce. Thankfully, I had the ability to perform a cost/benefit analysis. I could understand that a few moments of stolen pleasure would not be worth the destruction of my marriage. If I believed for a minute that Brenda would forgive me for being unfaithful, there's no doubt in my mind I would have engaged in multiple affairs.

I've come to believe my history with sexual promiscuity permanently damaged my ability to experience sex as God intended. I believe God designed sex to be a loving and lifelong bonding behavior to be experienced and enjoyed in the context of marriage. When a person experiences sex with multiple partners, I believe sex no longer creates the lifelong bond it was meant to establish. For decades, I believed the damage I experienced was mine alone until I read two articles while preparing to write this chapter. Since the restoration of your erectile function gives you new opportunities, I believe it's important to consider this information before you decide how you want to use your newfound abilities.

Oftentimes, those who preach sexual abstinence have been told to stop trying to impose their beliefs on others. But what if science could prove that sexual permissiveness does great damage to future sexual happiness? That's what Dr. Joe McIlhaney of the Medical Institute for Sexual Health in Austin says.

New research shows that sleeping around now could ruin your chances of having a happy, fulfilling marriage later. Research using brain scans shows powerful chemicals are released during sex that should create a powerful, everlasting bond. But that bond, which acts like adhesive tape or Velcro, is weakened when people break off with a sexual partner and moving from one to another to another. So, when it comes time to bond permanently with a spouse, the ability to bond is damaged. "The brain actually gets molded to not accept that deep emotional level that's so important for marriage.[3]

This doesn't mean postponing sex until marriage grants you some special ability to remain faithful without any struggle or temptation. However, according to McIlhaney, "When those who've been sexually promiscuous marry, they're more likely to have a divorce than people who were virgins when they got married."[4]

Science also affirms a destructive downside to casual sex. "Every time a person has sexual intercourse or intimate physical contact, bonding takes place. Whenever breakups occur in bonded relationships there is confusion and often pain in the brains of the young people involved because the bond has been broken."[5]

The problem described here also has a much deeper and long-lasting effect. The authors report: "Further, there is evidence that when this sex/bonding/breaking-up cycle is repeated a few, or many times—even when the bonding was short-lived—damage is done to the important, built-in ability to develop significant and meaningful connection to other human beings."6

I believe this research is accurate based on the damage I've known experientially and struggled with for decades. I've experienced additional damage, but I don't find it discussed in the scientific literature. If you've engaged in sex with multiple partners, at some point in your marriage, most, if not every one of your previous sexual partners, will make an appearance in your marital bed via your imagination. A few of your pre-marital partners will remain as uninvited guests who will refuse to leave your marital bed even when asked to do so. It took me decades to clear out all of those uninvited guests from my bedroom. Even now, thirty-five years into my marriage, I'll receive an unwelcome visitor in my mind who will distract me from enjoying my sexual experience with my wife.

This is the reason why I'm against sex therapists who recommend pornography for men and couples coping with ED. I was exposed to pornography for the first time in my early teens. For more than a decade, pornography was part of my life through movies, pictures, and print. Pornography burns images into your mind that can and will remain there, possibly for the rest of your life. You should decide before you use pornography whether or not you want to make the images you see part of your permanent sexual history. The decision to use pornography is a personal one. Based on my own experiences, I suggest reading a few of these articles before you decide to play with fire:

1. The Dangers of Internet Porn[7]
2. The effects of pornography on individuals, marriage, family and community[8]
3. How pornography influences and harms sexual behavior[9]
4. Porn-Induced Erectile Dsyfunction[9]

I can say with conviction that I was harmed by my promiscuity and my habitual use of pornography. You have the freedom to disagree. You can use pornography and engage in sex any time with any person of your choosing. I shared my experiences because our culture and our media are heavily invested in promoting both pornography and casual sex. I believe it's important for both single men and single women to understand there are unintended consequences for choosing to use your newfound erectile abilities to engage in sex with multiple partners. If you decide to engage in casual sexual sex with multiple partners or use pornography, I'm compelled to say, "Danger, Will Robinson!"

Questions to Consider:

1. What are your motivations for having sex?
2. How do you imagine a restoration of your erectile abilities will impact your self-esteem?

3. How will the restoration of your erectile abilities change your behavior on dates?
4. When and how will you tell a date that you have an implant?
5. Does the ability to maintain an erection when you want, and for as long as you want, give you a reason to feel superior in some way? If so, how will that effect your behavior in your relationships?

CHAPTER 19

Looking Back and Moving Forward

In my early and darkest days coping with erectile dysfunction, I felt so worthless and defective that if Brenda had said, "I can't tolerate being married to a man who will be impotent for the rest of his life; I want a divorce," I would have responded by saying, "I don't blame you. Let's divide our property amicably. I wish you well on your journey to find love with a fully-functioning man." After that, I would have withdrawn from my social life and lived out the rest of my days isolated and alone. In fact, that's exactly what I did within our marriage for the first eighteen months or so. It took me a long time to finish grieving what I had lost and a lot of patience and encouragement from Brenda for me to agree with her belief that we were still capable of maintaining an exciting sex life while I was impotent.

With time, effort, imagination, humor, and a lot of grace, we developed a mutually satisfying sexual relationship for two of the four years I was functioning occasionally but mostly impotent. For those men who choose to live with their impotence, it's important to understand you might give up on your sexuality as a result of getting stuck in the Diver phase where you feel hopeless, helpless, useless, and depressed.

That's the worst possible time to decide to live out the rest of your life without the joy, pleasure, or the physical and emotional connection that comes from sex. We live in an information age. Books, online support groups, and professionals can help move you out of the Diver phase of adjustment. If you find yourself stuck in the Diver phase, it's a costly mistake to cope with impotence alone. The decisions you make in the Diver phase are usually self-destructive and add to the misery of living with erectile dysfunction.

Even though Brenda and I discovered ways to enjoy our sexual relationship, I remained stuck in the Diver phase for years. I was surprised at the length of time it took me to finishing grieving the losses to my sexuality as a result of prostate surgery. When I was told my ED was permanent, I didn't know what to do next, but I knew I wasn't ready to give up that part of our sexual relationship for the rest of my life. I needed to find a way to reclaim what my prostate cancer treatment had taken away.

After losing my erectile ability for four years and then regaining it, I possess a newfound appreciation and gratitude for the restoration of my erectile function. From my perspective, at age sixty-four, I'm surprised to say that our sex life is better than it's ever been in the history of our thirty-five-year marriage. My decision to have an implant was one of the best treatment decisions I've made in my lifetime.

As I was writing this chapter I ran across an article online with the following headline: *While erectile dysfunction increases, use of penile implants* decline.[1] I find that trend tragic for men and couples. There is no reason to live without hope or to live out the rest of your life without any form of enjoyable sexual satisfaction. I've shared my experiences for the express purpose of making it clear that impotence does not doom men to a life they will hate. It's possible to experience an exciting sexual and romantic relationship living with erectile dysfunction. If you want your erectile function restored, you have a variety of treatment options from which to choose.

If you have lived with erectile dysfunction for a long time and have decided to have implant surgery, you might experience a common regret found among men with penile implants. You'll regret you waited so long before embarking on the journey toward an implant. To those men who feel like I did, that you don't want to live the rest of your life impotent, implant surgery will usher in a new chapter in your life that will allow you to say farewell to erectile dysfunction.

Looking back, I regret that I allowed shame, embarrassment, grief, depression, hopelessness, ignorance, and fear, to rob me of my ability and desire to make the best choices during each phase of my journey with erectile dysfunction. I regret every year I spent alone in the Diver phase, believing in the reality of my survival strategies, leaving me alienated from my loving and supportive wife. I regret we didn't seek professional help years earlier in our journey with ED. I wrongly assumed a therapist who wasn't impotent couldn't understand what I was going through.

Looking forward, it's time for me to retire the "I'm all alone in the world" and the "prepare to be deserted" survival strategies. Both of these survival strategies have caused nothing but trouble in our marriage. One year after surgery, I couldn't be happier with the decision to go through with implant surgery. We've enjoyed many romantic vacations together and plan to take a few long weekend romantic vacations each year. My sense of confidence in the bedroom is at the highest level it's ever been. Considering it was recently the lowest it's ever been, that's an amazing transformation.

All remnants of depression associated with living with ED are in the past. It was a tough journey, but it was worth all the pain and suffering to get where we are today. Brenda and I have restored both our emotional and sexual connections with each other. I feel penile implant surgery gave us back what treating my prostate cancer took away. There's no doubt in either of our minds that implant surgery was the right decision for us.

As you deal with the challenges of coping with ED, you have an important advantage that we didn't have as we began our journey. You have the opportunity to learn from our experience. I didn't know about the different ways to cope with ED. I had no idea it was important to take time to grieve the loss of my sexual triggers and discover creative ways to reestablish them.

You know that PTSD isn't limited to those who have served in the military, and you've been informed of the ways in which PTSD can make its presence known. It won't take you by surprise if you find yourself believing the lie that manhood equals an erect penis. You know the necessity of challenging that belief and that your faith can help you to reclaim your manhood. You've learned about a variety of options that are available to treat ED. You know there are five love languages you can learn.

You've discovered that depression and shame can prevent you from seeking help and destroy your motivation to discover new and creative ways to bring pleasure, sensuality, and romance into your relationship. You've been warned to watch for the emergence of survival strategies, which leave you stuck in the past and hamper your ability to live and learn from your present reality. You have an awareness of the destructive nature of mood-altering behaviors, such as using/abusing drugs, alcohol, computers, TV, pornography, or engaging in an extra-marital affairs or sex with multiple partners.

You have some ways to motivate a resistant partner to seek out professional consultations. You've learned about three personality traits that can help you to meet the challenges of living with ED. If you apply the lessons we've shared to your circumstances, you can avoid years, possibly a lifetime, of emotional and relational misery. You can pick yourself up and get back on the path toward developing a new sexuality based on how you decide to live with and/or treat your erectile dysfunction. For your sake, I hope you are not strong enough to accomplish it alone. Discovering you need a team allows you to develop strength through weakness, which is the most effective strength you can possess.

I've saved best until last. Here's a powerful, relationship-changing behavior that's easily doable. Men with ED typically withdraw from all forms of physical affection. Avoiding physical touch negatively impacts your mental health as well as the mental health of their partner. You feel as though you are living together alone. The importance of touch and physical affection cannot be overstated.

Studies have shown that people need and crave being touched more than they desire sex, money or social status. Basic physical (non-sexual) affection such as holding hands, getting a massage, being caressed and getting hugged are very important to our species and sometimes, people who go without being touched in an affectionate way for long periods of time often become depressed, they lose their interest in daily life activities and tend to isolate themselves from people.[2]

Here's the solution, from the research of world-renowned relationship experts John and Julie Gottman:

"Hold her. Hold her before sex, during sex and after sex. Hold her when you're dating. Hold her when you married. Hold her when she's upset. Hold her when she's happy. Hold her when she's scared. Hold her when she feels unworthy of being held, and hold her when she's mad. Hold her every time she needs to be held and you will be her best lover. It's as simple as that."[3]

In addition to taking time to hold your partner, make sure to kiss your partner every day. Make a few of those kisses last longer than six seconds. That's how long it takes to release the hormones that can impact the way you feel about your partner. There are psychological, relational, sexual, and hormonal benefits to kissing your partner every day.[4]

When we kiss, our levels of oxytocin skyrocket, so kissing does make us feel closer. Kissing also stimulates nerve endings on our lips, which sparks the release of dopamine. Dopamine is a neurotransmitter that is active in brain circuits associated with pleasure, and it makes us feel happy. Dopamine is also associated with feeling rewarded, which can make us want to repeat a behavior, so one kiss with someone really can lead you to wanting more. Endorphins are released when we kiss, neuropeptides that gives us a happy buzz, like after we exercise. Phenylethylamine levels also increase during a kiss. This chemical is actually similar to amphetamines. Not only does it make us happy, it has aphrodisiac effects, too!

If you use affectionate physical touch to hold your partner and kiss your partner each and every day, both of you will reap the individual and relational rewards and benefits whether or not you have penile implant surgery.

As you come to the end of this book, you have some life-altering decisions to make. There's one unhealthy option I hope you'll reject categorically. That's the option of doing nothing, which sentences you and your partner to a lifetime of living with misery and depression. By now, you've discovered a host of positive options with far better outcomes. Here's a few of them:

1. You can get a physical exam to determine the underlying cause for your ED.
2. You can begin to explore which treatment option is best for you.
3. If you decide the best option involves living with ED, you can develop an expertise in enjoying non-erection-based sex.
4. If medication or injections are effective, will be able to restore your sex life
5. If you choose implant surgery, this book can help you in your journey.

Going through this entire journey with prostate cancer, impotence, and implant surgery was highly stressful. Both Brenda and I suffered individually and as a couple. We've both learned some valuable life lessons in these seasons of our life together. While implant surgery is not for everyone, If you've tried all the treatment options available and they've all failed and it's still important for you to regain your erectile function, consider implant surgery.

One year after surgery, Brenda and I are having the time of our lives. As I type these words, we are preparing for our first romantic vacation of 2016. Unlike many of our romantic vacations before my implant surgery, a few nasty vacation crashers will not be joining us on our getaway. I've said farewell to headaches, facial flushing, backaches, and every other annoying side effect from ED medications. My other vacation crasher, worry, can no longer rob me of my joy. Two other vacation crashers, disappointment and failure, can no longer mess with my confidence or manhood. They're all banished from my vacations, my bedroom, and my life.

There's no doubt in either of our minds regarding my ability to rise to every occasion. I've finally put an end to the silence, sadness, suffering, and shame in my life. Now that you've read about our experience, you have the option and the tools to end the silence, sadness, suffering, and the shame in your life as well.

It's our prayer that you'll find new ways to enjoy your sexuality with or without a penile implant. In either case, you can enjoy romance, sensual touch, and experience an exciting sex life together

If at any point in your journey you'd like to contact either of us or inquire about our availability for a speaking engagement, you can reach us at our website, HYPERLINK "http://whereisyourprostate.com/" or send us an e-mail at copingwithed@gmail.com

You can also visit us on Facebook "HYPERLINK" https://www.facebook.com/whereisyourprostate/"

Check out our ED and Penile Implant forum at "HYPERLINK" "http://penileimplants.proboards.com/" http://penileimplants.proboards.com"

Blessings to you in whatever you decide,

Rick & Brenda Redner

APPENDIX

Frequently Asked Questions

(Answered by Rick)

1. **Knowing what you know today would you choose implant surgery again?** Positively and without hesitation, yes!
2. **How has implant surgery changed your life?** For the better! My confidence in the bedroom is at the highest it's ever been. Whether or not I use the implant, I feel grateful every day of my life that my wife and I can resume the joy, pleasure, and closeness we experience during intercourse. Implant surgery gave me what prostate cancer took away. Saying goodbye to impotence remains the favorite goodbye I've experienced in my lifetime.
3. **Does the implant change your sensation?** My penis feels different because of the cylinders, one on each side of my penis. It takes more effort for me to achieve an orgasm, but the sensations I feel during intercourse are just as pleasurable now as they were prior to implant surgery. I also find my orgasms are more intense now than when I experienced them with a flaccid penis.
4. **Will my insurance pay for the procedure?** If your ED has a physical/organic cause, most insurance companies, including Medicare, will pay for the implant procedure. Depending upon your policy, you might have no out-of-pocket costs or you might, as I did, have thousands of dollars of out-of-pocket costs.

5. **How long will the implant function?** Most implants work properly for ten to fifteen years or more.
6. **Will my erections be as large as they were prior to surgery?** The loss of length is the most common complaint men express after surgery. Expandable devices are available that will allow the penis to stretch to its full potential. Speak to your surgeon about the possibility of that type of implant if post-surgery erectile size is an important issue for you.
7. **What are the risks of surgery?** As with any surgery, risks are involved in penile implant surgery. Here are a few: internal bleeding, post-surgery infection, implant device failure, and internal erosion, which involves the implant wearing away the skin from inside the penis. To eliminate the risk of an allergic reaction to drugs, make sure you tell your surgeon about any drug allergies *before* surgery.
8. **How long will it take to recover from surgery?** Most men can return to work within three weeks of surgery
9. **How long must I wait before my implant is activated?** Most surgeons recommend waiting six weeks after surgery.
10. **Is the implant noticeable?** No one will know you have an implant. You might find your non-erect penile size to be larger after surgery due to the cylinders that are placed in your penis.
11. **Is an implant uncomfortable?** For the first few months it felt uncomfortable to have a pump in my scrotum. Now I'm usually unaware of the presence of my implant. I did purchase underwear with scrotal support. I believe that adds to my comfort.
12. **Will an implant increase my desire for sex?** If depression or avoiding failure impacted your desire to have sex, those issues will no longer inhibit you, but the implant itself does nothing to increase your libido.
13. **Can Peyronie's disease be treated with an implant?** Yes, it is possible the curvature of the penis can be corrected with an implant.
14. **I'm twenty-five years old. Will an implant make it impossible for me to have children?** No. Penile implant surgery does not prevent sperm production or stop ejaculation.

15. **I'm seventy years old. Am I too old for an implant?** Not at all! Men in their nineties who are healthy enough for surgery have had success with an implant.
16. **Will my implant cause problems during airport screening?** No, it won't. The new body scanners allow TSA employees to see your implant, but it will not cause any issues with you getting through airport screening.
17. **Will the implant improve my relationship with my partner?** Having an implant does nothing to resolve the emotional issues and conflicts you experience in your relationship. In fact, it's possible your newfound abilities will increase the level of conflict in your relationship rather than diminish them. The reality is that your implant alone will not provide you with a happily-ever-after relationship.
18. **I don't have the insurance or cash to for this procedure. Is there an option for me?** Yes, there is! Go to this link: HYPERLINK "https://surgeo.com/penile-implant". You'll find information about implant surgery packages you can purchase or finance.
19. **Is there a way to speak to men who have an implant?** An online service called Patient Perspectives is available. Here's a link to their website: HYPERLINK "https://www.patientperspectives.org/"
20. **If I choose to live with erectile dysfunction, am I doomed to lonely, single life?** Absolutely not! However, there are people for whom impotence is a dealbreaker. Those folks will have no interest in a romantic relationship with you, andthat's okay. The erection-focused needs of the few do not disqualify you from dating or getting married. If you are kind, affectionate, loving, generous, considerate, trustworthy, fun loving, and possess a proficiency in speaking all five love languages, you'll be considered a great catch! There's no reason for you to avoid dating or marriage. If you're struggling with a loss of confidence or depression, I suggest you deal with those issues before you begin dating.

GLOSSARY

Activation day: the official day your surgeon gives you permission to use your penile implant. Usually, you are given an appointment four to six weeks after your surgery, on which you will learn how to inflate and deflate your implant.

Catheters: hollow, partially flexible tubes that collect urine from the bladder

Climacturia: the involuntary release of urine at the time of orgasm during sexual activity

Cycling the pump: the act of inflating and then deflating the penile implant

Depression: a mood disorder marked by sadness, inactivity, difficulty thinking and concentrating, a significant increase or decrease in appetite and time spent sleeping, and feelings of dejection and hopelessness. Sometimes accompanied by suicidal thoughts.

Bilateral Nerve-sparing Surgery: During the surgical removal of the prostate gland, an attempt is made to spare the two cavernous nerve sheaths, which preserve erectile function.

Impotence: chronic inability to attain or sustain an erection for intercourse

Oxygenated blood: Oxygen-rich blood is one of the most important components for erectile health. Oxygen levels vary widely from reduced levels in the flaccid state to high levels in the erect state. During sleep, most men have three to five erections per night, bringing oxygen-rich blood to the penis.

Penile Rehabilitation: a program designed to improve blood flow to the penis to maximize the chances of regaining full erectile function. Medication, a vacuum pump, and/or penile injections are part of this program.

Prostate cancer: a disease that affects men only. Cancer begins to grow in the prostate, a gland that is part of the male reproductive system. Approximately fourteen percent of men receive this diagnosis in their lifetime.

Post-traumatic stress disorder (PTSD): a disorder that develops in some people who have witnessed or experienced a shocking, frightening, dangerous, or life-threatening event

Robotic prostatectomies: a surgery where the surgeon sits at a console using instruments attached to a mechanical device (robot).

Scrotal Support: an undergarment that provides testicular support/

Sexual Triggers: any sight, sound, touch, or smell that leads to an erection

Urinary retention: the inability to completely or partially empty the bladder

Venous leak: when the veins can't keep blood in the penis during an erection. The leak prevents a man from maintaining an erection.

NOTES

Introduction

[1] Kathleen Doheny. “Most Men With ED Don’t Seem to Get Treatment.” WebMD. *WebMD.* http://www.webmd.com/erectile-dysfunction/news/20130506/most-men-with-erectile-dysfunction-dont-seem-to-get-treatment.

[2] Mayo Clinic staff. “Erectile Dysfunction: Causes.” *Mayoclinic.org.* HYPERLINK “http://www.mayoclinic.org/diseases-conditions/erectile-dysfunction/basics/causes/con-20034244”

Chapter 1: Losing Your Erectile Abilities—A Traumatic Life-changing Event

[1] “Erectile Dysfunction.” *National Institute of Diabetes and Digestive and Kidney Diseases.* http://www.niddk.nih.gov/health-information/health-topics/urologic-disease/erectile-dysfunction/Pages/facts.aspx.

[2] Tara Parker-Pope. August 27, 2008. “Regrets After Prostate Surgery.” *New York Times.* http://well.blogs.nytimes.com/2008/08/27/regrets-after-prostate-surgery/?_r=0.

[3] Redner, Rick, and Redner, Brenda. 2013. *I Left My Prostate in San Francisco: Coping with the Emotional, Relational, Sexual, and Spiritual Aspects of Prostate Cancer.* Modesto: WestBow Press.

[4] Print.Psychosocial and Relationship Issues in Men With Erectile Dysfunction.” Medscape Log In. Web. 05 Mar. 2016. http://www.medscape.com/viewarticle/551562_5>.

[5] Patrick J. DeMeo. 2006. “Psychosocial and Relationship Issues in Men With Erectile Dysfunction.” *Medscape.* http://www.medscape.com/viewarticle/551562.

[6] “Depression in Men: Why It’s Hard to Recognize and What Helps.” *HelpGuide.org.* HYPERLINK “http://www.helpguide.org/articles/depression/depression-in-men.htm” http://www.helpguide.org/articles/depression/depression-in-men.htm.

[7] Laurence Roy Stains. January 28, 2010. “I Want My Prostate Back.” *Men’s Health.* http://www.menshealth.com/health/coping-with-prostate-cancer.

[8] "Consequences of Erectile Dysfunction." *Sexual Health Australia*. http://www.sexualhealthaustralia.com.au/consequences_of_ed.html.

[9] Mark Henricks. "What is Erectile Dysfunction (ED)?" *EverydayHealth.com*. HYPERLINK http://www.everydayhealth.com/erectile-dysfunction/who-gets-erectile-dysfunction.aspx

[10] "Depression in Men: Why It's Hard to Recognize and What Helps." *HelpGuide.org*. HYPERLINK "http://www.helpguide.org/articles/depression/depression-in-men.htm" http://www.helpguide.org/articles/depression/depression-in-men.htm.

[11] John M. Grohol. "Where to Get Help for Depression." *Psych Central*. HYPERLINK "http://psychcentral.com/lib/where-to-get-help-for-depression/" http://psychcentral.com/lib/where-to-get-help-for-depression/.

[12] "Find a Therapist." *Anxiety and Depression Association of America*. HYPERLINK "http://treatment.adaa.org/" http://treatment.adaa.org/.

[13] Depression Assessment: How Well Are You Managing Symptoms?" *WebMD*. HYPERLINK "https://www.webmd.com/depression/depression-assessment/default.htm" https://www.webmd.com/depression/depression-assessment/default.htm.

Chapter 2: ED is a Thief

[1] "Depression Caused by Erectile Dysfunction: Symptoms and Treatments." *WebMD*. HYPERLINK "http://www.webmd.com/erectile-dysfunction/ed-related-depression" http://www.webmd.com/erectile-dysfunction/ed-related-depression.

[2] Consequences of Erectile Dysfunction.", Impotence. Web. 12 Mar. 2016. http://www.sexualhealthaustralia.com.au/consequences_of_ed.html.

Chapter 3: How You Think About ED Changes Everything

[1] Ian Golding. May 7, 2015. "Thrivers, Survivors and Nose-Divers! How to Help People Believe in Transforming Your Customer Experience?" *IJ Golding*. http://www.ijgolding.com/2015/05/07/thrivers-survivors-and-nose-divers-how-to-help-people-believe-in-transforming-your-customer-experience/.

[2] "Suicide Prevention." *National Institute of Mental Health*. HYPERLINK "https://www.nimh.nih.gov/health/topics/suicide-prevention/index.shtml" https://www.nimh.nih.gov/health/topics/suicide-prevention/index.shtml.

Chapter 4: What You Don't Know Can Hurt You

[1] "Erectile Dysfunction: What NOT to Say When He Can't (cough) Perform." *VibrantNation.com*. http://www.vibrantnation.com/erectile-dysfunction-not-say-cant-cough-perform/.

[2] Lisa Zamosky. "12 Signs of Depression in Men." *Health.com*. http://www.health.com/health/gallery/0,,20521449,00.html.

[3] Healthline editorial team. December 12, 2013. "How Can I Get Help for Depression?" *Healthline.* http://www.healthline.com/health/depression/help-for-depression#AlternativeandComplementaryTherapies1.

Chapter 5: "Danger, Will Robinson!"

[1] FrankTalk ED Support Groups: HYPERLINK "http://www.franktalk.org/phpBB3/" http://www.franktalk.org/phpBB3/

[2] MD Junction ED Support Group: HYPERLINK "http://www.mdjunction.com/erectile-dysfunction" http://www.mdjunction.com/erectile-dysfunction

[3] Daily Strength ED Support Group: HYPERLINK "http://www.dailystrength.org/c/Impotence-Erectile-Dysfunction/support-group" http://www.dailystrength.org/c/Impotence-Erectile-Dysfunction/support-group

[4] Erectile Dysfunction Support Group: HYPERLINK "http://edforum.org/" http://edforum.org/

[5] Rick's ED Support Group: HYPERLINK "http://penileimplants.proboards.com/" http://penileimplants.proboards.com

Chapter 6: What's Faith God to Do With It?

[1] New King James Bible

[2] Ibid

[3] Alterowitz, Ralph, Alterowitz, Barbara. 1999. *The Lovin' Ain't Over: The Couple's Guide to Better Sex after Prostate Disease.* Westbury, NY: Health Education Literary.

[4] Chapman, Gary D. 1995. *The Five Love Languages: How to Express Heartfelt Commitment to Your Mate.* Chicago: Northfield Publishing.

[5] "Discover Your Love Language." *The 5 Love Languages.* http://www.5lovelanguages.com/.

[6] "Need Some Advice? We can make it work again." *FrankTalk.org* HYPERLINK "http://www.franktalk.org/phpBB3/" http://www.franktalk.org/phpBB3/.

[7] "Impotence & Erectile Dysfunction Support Group." *Daily Strength.* http://www.dailystrength.org/c/Impotence-Erectile-Dysfunction/forum.

[8] "Erectile Dysfunction Community." *MedHelp.* http://www.medhelp.org/forums/Erectile-Dysfunction/show/124.

Chapter 7: The Long & Winding Road Toward Surgery

[1] Memorial Sloan Kettering. "Prostate Cancer & Sexual Health | Erectile Dysfunction | Memorial Sloan Kettering." *YouTube,* 18:32, https://www.youtube.com/watch?v=OQlGVT7HiF8.

[2] Mulhall, John P. 2008. *Saving Your Sex Life: A Guide for Men with Prostate Cancer.* Munster, IN: Hilton Publishing.

3 "What Does a Penile Rehabilitation Program (PRP) Involve?" What Does a Penile Rehabilitation Program (PRP) Involve?" *International Society for Sexual Medicine.* http://www.issm.info/education-for-all/sexual-health-qa/what-does-a-penile-rehabilitation-program-prp-involve.

4 "Inflatable Penile Prostheses and Satisfaction." *International Society for Sexual Medicine.* http://www.issm.info/news/sex-health-headlines/inflatable-penile-prostheses-and-satisfaction.

5 "Penile Implants" *Coloplast.* http://www.coloplastmenshealth.com/treatments/erectile-dysfunction-treatment/penile-implants/.

6 "Treating Erectile Dysfunction with Penile Implants." *Harvard Medical School.* http://www.harvardprostateknowledge.org/treating-erectile-dysfunction-with-penile-implants.

7 "89 Percent of Men with Penile Implants Are Able to Have Sex Says New Study." *Psych Central.* http://psychcentral.com/news/archives/2006-01/bpl-8pc010906.html.

8 "Erectile Dysfunction and Penile Prosthesis." *WebMD.* HYPERLINK "http://www.m.webmd.com/a-to-z-guides/penile-prosthesis" http://www.m.webmd.com/a-to-z-guides/penile-prosthesis.

9 Inflatable Penile Prostheses and Satisfaction." *International Society for Sexual Medicine.* HYPERLINK "http://www.issm.info/news/sex-health-headlines/inflatable-penile-prostheses-and-satisfaction" http://www.issm.info/news/sex-health-headlines/inflatable-penile-prostheses-and-satisfaction.

10 Beutler, LE, et al. December 1984. "Women's satisfaction with partner's penile implant. Inflatable vs. noninflatable prosthesis." *Urology.* 24(6) 552–558. http://www.ncbi.nlm.nih.gov/pubmed/6506394.

11 "Implants." *FrankTalk.org.* HYPERLINK "http://www.franktalk.org/phpBB3/viewforum.php?f=6" http://www.franktalk.org/phpBB3/viewforum.php?f=6.

Chapter 8: How to Choose Your Surgeon

1 "How to Choose a Doctor." *WebMD.* http://www.webmd.com/health-insurance/insurance-basics/how-to-choose-a-doctor.

2 "Consumer Reports Suggests Smart Ways to Choose a Surgeon." April 27, 2010. *Washington Post.* http://www.washingtonpost.com/wp-dyn/content/article/2010/04/26/AR2010042603382.html.

3 "Health Savings Account Frequently Asked Questions." *Key Bank.* https://www.key.com/personal/savings/health-savings-account-faq.jsp.

4 "Size Matters: What to Expect After Penile Implant Surgery." July 1, 2014. *Marketwired.* http://www.marketwired.com/press-release/size-matters-what-to-expect-after-penile-implant-surgery-1926153.htm.

5 Lee, King Chien Joe, Brock, Gerald B. March 2013. "Strategies for Maintaining Penile Size following Penile Implant." *Translational Andrology and Urology.* Vol. 2 No. 1. http://tau.amegroups.com/article/view/1459/2434.

6 "Local U Resources for Penile Implants." *MedicineNet.* HYPERLINK "http://www.medicinenet.com/penile_implants/city.htm" http://www.medicinenet.com/penile_implants/city.htm.

Chapter 9: Three Implant Options

1 "Types of Penile Implants." Advanced Urological Pare PC. https://www.urologicalcare.com/penile-implants-prosthesis/types-of-penile-implants/.
2 Komaroff, Dr. Anthony K. April 9, 2014. "What Are the Different Options of Penile Implants?" Web. 26 Feb. 2016. HYPERLINK "http://www.askdoctork.com/different-options-penile-implants-201404096260" http://www.askdoctork.com/different-options-penile-implants-201404096260.
3 "Surgical Treatment of Erectile Dysfunction." *EMedicineHealth.* http://www.emedicinehealth.com/surgical_treatment_of_erectile_dysfunction/page3_em.htm.
4 "Penile Prosthesis: Implant, Surgery, Effectiveness, and Sex Implications." *WebMD.* http://www.webmd.com/erectile-dysfunction/guide/penile-prosthesis.
5 "Conditions: Erectile Dysfunction." *Sex Health Matters.* http://www.sexhealthmatters.org/erectile-dysfunction/penile-implants-erectile-dysfunction/single.

Chapter 10: PTSD and Me—and Brenda, Too!

1 Scott, Cameron. March 12, 2015. "Cancer Treatment Leaves Survivors with PTSD Scars." *Healthline.* http://www.healthline.com/health-news/cancer-treatment-leaves-survivors-with-ptsd-scars-031215.
2 Frederik Joelving. Oct. 12, 2011. "Many Cancer Survivors Struggle with PTSD Symptoms." *Reuters.* http://www.reuters.com/article/2011/10/12/us-cancer-ptsd-idUSTRE79B7FT20111012.
3 Post-Traumatic Stress Disorder Symptoms, Causes and Effects." *PsychGuides.com.* http://www.psychguides.com/guides/post-traumatic-stress-disorder-symptoms-causes-and-effects/.
4 Mayo Clinic Staff. "Post-traumatic Stress Disorder (PTSD)." *Mayo Clinic.* http://www.mayoclinic.org/diseases-conditions/post-traumatic-stress-disorder/basics/risk-factors/con-20022540.
5 "Post-traumatic Stress Disorder." *Perelman School of Medicine.* http://www.med.upenn.edu/ctsa/ptsd_symptoms.htm.
6 "Post-Traumatic Stress Disorder and Cancer." January 2016. *Cancer.Net.* http://www.cancer.net/survivorship/life-after-cancer/post-traumatic-stress-disorder-and-cancer.
7 Rapaport, Lisa. April 21, 2015. "Sexual Dysfunction May Accompany PTSD." *Reuters.* http://www.reuters.com/article/us-ptsd-sex-mental-idUSKBN0NC28N20150421.

[8] "PTSD: Symptoms, Self-Help, and Treatment." *HelpGuide.org.* http://www.helpguide.org/articles/ptsd-trauma/post-traumatic-stress-disorder.htm.

[9] "PTSD: National Center for PTSD." *US Department of Veterans Affairs.* HYPERLINK "http://www.ptsd.va.gov/public/treatment/therapy-med/treatment-ptsd.asp" http://www.ptsd.va.gov/public/treatment/therapy-med/treatment-ptsd.asp.

Chapter 11: Check Your Unrealistic Expectations At The Door

[1] "Treating Erectile Dysfunction with Penile Implants." *Harvard Medical School.* http://www.harvardprostateknowledge.org/treating-erectile-dysfunction-with-penile-implants.

[2] "Inflatable Penile Prostheses and Satisfaction." *International Society for Sexual Medicine.* http://www.issm.info/news/sex-health-headlines/inflatable-penile-prostheses-and-satisfaction.

[3] "What Is the Success Rate for Penile Implants?" *Sharecare.com.* https://www.sharecare.com/health/erectile-dysfunction-treatment/what-success-rate-penile-implants.

Chapter 12: Before Implant Surgery—A Wife's Perspective

[1] "A Woman's 6-step Survival Guide for Male Impotency." *Phoenix.org.* http://www.phoenix5.org/companions/6steps.html.

[2] Worth, Tammy. May 10, 2012. "What She Really Thinks of Your Penis." *Men's Health.* http://www.menshealth.com/sex-women/erection-returns.

[3] Wait, Marianne. "A Partner's Guide to Erectile Dysfunction." *WebMD.* http://www.webmd.com/erectile-dysfunction/features/a-womans-guide-to-ed.

[4] "A Woman's Perspective on Erectile Dysfunction." *Total Performance Medical Center.* http://www.totalperformancemedical.com/womans-perspective-erectile-dysfunction.

[5] Stewart, Kristen. "How Women Can Deal with Erectile Dysfunction." *EverydayHealth.com.* http://www.everydayhealth.com/erectile-dysfunction/how-women-can-deal-with-erectile-dysfunction.aspx.

Chapter 13: Implant Surgery Here I Come

[1] Roth, H, et al. 2012. "Pain Prevents The Early Activation Of Inflatable Penile Prostheses." *The Internet Journal of Urology.* Vol. 9, No. 4. http://ispub.com/IJU/9/4/14361.

[2] Mayo Clinic Staff. "Penile Implants: How You Prepare." *Mayo Clinic.* http://www.mayoclinic.org/tests-procedures/penile-implants/basics/how-you-prepare/prc-20013140.

[3] "How Should I Prepare for Penile Implant Surgery?" *Sharecare.* HYPERLINK "https://www.sharecare.com/health/erectile-dysfunction-treatment/how-prepare-penile-implant-surgery" https://www.sharecare.com/health/erectile-dysfunction-treatment/how-prepare-penile-implant-surgery.

Chapter 14: Using—Then Abusing—Pain Medication

[1] "Drug overdose deaths in the United States hit record numbers in 2014." *Centers for Disease Control and Prevention.* http://www.cdc.gov/drugoverdose/epidemic/.

[2] "Oxycodone Abuse Side Effects, Addiction Signs & Symptoms." Delta Medical Center. HYPERLINK "http://www.deltamedcenter.com/addiction/prescription-drugs/oxycodone/effects-symptoms-signs" http://www.deltamedcenter.com/addiction/prescription-drugs/oxycodone/effects-symptoms-signs.

Chapter 15: Activation Day

[1] Roth, H, et al. 2012. "Pain Prevents The Early Activation Of Inflatable Penile Prostheses." *The Internet Journal of Urology.* Vol. 9, No. 4. http://ispub.com/IJU/9/4/14361.

[2] Ibid

[3] "Demo of Patient with Penile Implant in the Supine Position." *YouTube,* 10:47, HYPERLINK "https://www.youtube.com/watch?v=ESg0Bd2zLFg" https://www.youtube.com/watch?v=ESg0Bd2zLFg.

Chapter 16: Sometimes Survival Isn't the Best Choice

[1] "Trauma: How to Move Beyond Survival Mode." *Promises Treatment Centers.* https://www.promises.com/articles/mental-health/trauma-move-beyond-survival-mode/.

[2] "Getting Triggered." *1in6.org.* https://1in6.org/men/get-information/online-readings/self-regulation-and-addictions/getting-triggered/.

[3] Worth, Tammy. May 10, 2012. "What She Really Thinks of Your Penis." *Men's Health.* http://www.menshealth.com/sex-women/erection-returns.

[4] Castleman, Michael. "Great Sex Without Intercourse." *AARP.* HYPERLINK "http://www.aarp.org/home-family/sex-intimacy/info-12-2012/great-sex-without-intercourse.html" http://www.aarp.org/home-family/sex-intimacy/info-12-2012/great-sex-without-intercourse.html.

[5] "A Woman's 6-step Survival Guide for Male Impotency." *Phoenix.org.* http://www.phoenix5.org/companions/6steps.html.

[6] Case-Lo, Christine. July 10, 2013. "Tips for Sexually Frustrated Couples." *Healthline.* http://www.healthline.com/health-slideshow/tips-sexually-frustrated-couples.

[7] "Readjusting after Erectile Dysfunction." *Canoe.com.* http://chealth.canoe.com/Channel/Sexual-Health/Your-sex-life/Readjusting-after-erectile-dysfunction.

Chapter 17: Post Implant Surgery—A Wife's Perspective

1. Beutler, LE, et al. December 1984. "Women's Satisfaction with Partners' Penile Implant. Inflatable vs. Noninflatable Prosthesis." *Urology.* 24(6) 552–558. http://www.ncbi.nlm.nih.gov/m/pubmed/6506394/.
2. "What Is the Success Rate for Penile Implants?" *Sharecare.com.* https://www.sharecare.com/health/erectile-dysfunction-treatment/what-success-rate-penile-implants.
3. Beutler, LE, et al. December 1984. "Women's Satisfaction with Partners' Penile Implant. Inflatable vs. Noninflatable Prosthesis." *Urology.* 24(6) 552–558. http://www.ncbi.nlm.nih.gov/m/pubmed/6506394/.
4. "A wife's perspective." *Phoenix5.com.* http://www.phoenix5.org/sexaids/implants/implantPersWendy.html.
5. Beutler, LE, et al. December 1984. "Women's Satisfaction with Partners' Penile Implant. Inflatable vs. Noninflatable Prosthesis." *Urology.* 24(6) 552–558. http://www.ncbi.nlm.nih.gov/m/pubmed/6506394/.
6. "Testimonials." *Coloplast.* HYPERLINK "http://www.coloplastmenshealth.com/patient-stories/" http://www.coloplastmenshealth.com/patient-stories/.

Chapter 18: Single With An Implant

1. Allison. May 26, 2014. "Candy Spelling Says She Once Broke Up With A Guy Because Of His Bionic Penis." *Dlisted.com.* http://dlisted.com/2014/05/26/candy-spelling-says-she-once-broke-up-with-a-guy-because-of-his-bionic-penis/.
2. Miller, Kelli. "The Top 20 Reasons Why People Have Sex." *WebMD.* http://www.webmd.com/sex-relationships/guide/why-people-have-sex.
3. Strand, Paul. December 30, 2010. "Sexually Indulgent Now, Marriage Ruined Later?" *CBN News.* http://www.cbn.com/cbnnews/healthscience/2010/march/sexually-indulgent-now-marriage-ruined-later/?mobile=false.
4. Ibid.
5. Leap, Dennis. February 2011. "The Emotional Corrosion of Casual Sex." *TheTrumpet.com.* https://www.thetrumpet.com/article/7750.24.0.0/the-emotional-corrosion-of-casual-sex.

Chapter 19: Looking Back and Moving Forward

1. Smart, Ben. June 23, 2015. "While ED Increases, Use of Penile Implants Declines." *CNN.* HYPERLINK "http://www.cnn.com/2015/06/23/health/penile-implants-erectile-dysfunction/" http://www.cnn.com/2015/06/23/health/penile-implants-erectile-dysfunction/.
2. "Craving Touch/physical Affection Is an Important Part of Being Human... (depressed, Girlfriend)" *City-Data.com.* http://www.city-data.com/forum/psychology/1780367-craving-touch-physical-affection-important-part.html.

[3] Gottman, John Mordechai, et al. 2016. *The Man's Guide to Women.* New York: Rodale Books, 121.

[4] "The Science behind Kissing: 10 Things That Happen When We Kiss." June 21, 2015. http://sparklyscience.com/2015/06/21/the-science-behind-kissing-10-things-that-happenwhen-we-kiss/.

[5] Redner, Richard. "Erectile Dysfunction & Penile Support Forum." HYPERLINK "http://penileimplants.proboards.com/" http://penileimplants.proboards.com

[6] Redner, Rick, and Redner, Brenda. 2013. *I Left My Prostate in San Francisco--Where's Yours? Coping with the Emotional, Relational, Sexual, and Spiritual Aspects of Prostate Cancer.* Modesto: WestBow Press.

Appendix
Frequently Asked Questions

[1] Persaud, Natasha. December 8, 2015. "Why Men Are Satisfied With Penile Implants." *Renal and Urology News.* http://www.renalandurologynews.com/erectile-dysfunction-ed/why-men-are-satisfied-with-penile-implants/article/458374/.

[2] "Answers For Men—Understanding Your Treatment Options." *American Medical Systems.* http://www.tucc.com/UserFiles/file/AMS Answers ED.pdf.

[3] Elist, James J. "Penile Implant Sensation, What Is Sex like with the Penile Implant?" *Drelist.com.* http://www.drelist.com/penile-implant-sensation/.

[4] "Insurance Information." *Boston Scientific.* https://americanmedicalsystems.com/en/patients/insurance.html.

[5] Atkinson, George. January 12, 2006. "Penile Implants The "Forgotten" Solution For Erectile Dysfunction." *Altpenis.com.* http://www.altpenis.com/news/20060012013226data_trunc_sys.shtml.

[6] Eigner, EB, et al. 1991. "Penile Implants in the Treatment of Peyronie's Disease." *Journal of Urology.* http://www.ncbi.nlm.nih.gov/m/pubmed/1984103/.

[7] Garber, Bruce B. "Overcoming Impotence." *Garber-online.org.* http://www.garber-online.com/penile-implants-and-impotence.htm.

[8] "Size Matters: What to Expect After Penile Implant Surgery." July 1, 2014. *Marketwired.com.* HYPERLINK "http://m.marketwired.com/press-release/size-matters-what-to-expect-after-penile-implant-surgery-1926153.htm" http://m.marketwired.com/press-release/size-matters-what-to-expect-after-penile-implant-surgery-1926153.htm.

[9] Silverberg, Cory. May 18, 2016. "Erectile Dysfunction and Dating." *About.com.* http://sexuality.about.com/od/Erectile-Dysfunction/a/Erectile-Dysfunction-And-Dating.htm.

[10] Bacher, Renee. "Dating and Erectile Dysfunction (ED)." *WebMD.* http://www.webmd.com/men/features/dating-and-ed.

[11] "5 Ways to Deal with Impotence in a Dating Relationship." *The Brothers Media*. http://www.thebrothersnetwork.net/diabetes/2013/11/deal-with-impotence-in-a-dating-relationship.html.

[12] "Penile Implants – Frequently Asked Questions." *University of Utah Health Care*. http://healthcare.utah.edu/urology/sexual-dysfunction/penile-implants/faqs.php.

[13] "Erectile Dysfunction: Penile Prosthesis." *MedicineNet.com*. http://www.medicinenet.com/script/main/mobileart.asp?articlekey=42905.

[14] "Penile Implant." *Surgeo*. https://surgeo.com/penile-implant.

[15] "AMS Champion Program." *Patient Perspectives.org*. https://www.patientperspectives.org/.

INDEX

F

G

H

I

L

M

N

O

P

R

S

T

U

V

W

Lightning Source UK Ltd.
Milton Keynes UK
UKOW02f0823141116
287566UK00001B/107/P

9 781483 453903